McDougall, Bruce, 1950-
Ted Rogers

Includes bibliographical references (p. [200]-205) and
index.
721114   ISBN:1896176089

1. Rogers, Ted, 1933- 2. Rogers Telecommunications
Limited. 3. Businessmen - Canada - Biography. 4.
                    (SEE NEXT CARD)
278   95NOV08   3597/   1-403394

Ted Rogers

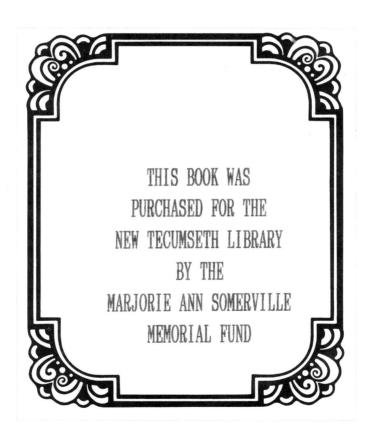

Ted Rogers

Bruce McDougall

BURGHER
BOOKS

Burgher Books
10 Edmund Avenue
Toronto, Ontario
M4V 1H3 Canada

Canadian Cataloguing in Publication Data
McDougall, Bruce
        Ted Rogers: an executive profile
Includes bibliographical references and index.
ISBN 1-896176-08-9
1. Rogers, Ted, 1933– . 2. Rogers Telecommunications Limited.
3. Businessmen – Canada – Biography. 4. Telecommunications – Canada – Biography.
I. Title.
TK5243.R6M33 1995   384.5'092   C95-932082-2

Distributed by:
Raincoast Books Limited
8680 Cambie Street
Vancouver, B.C.
V6P 6M9 Canada
Toll-free order line: 1-800-663-5714

Editor: Meg Taylor
Text design: Counterpunch/Linda Gustafson

Printed and bound in Canada
95 96 97 98 99   5 4 3 2 1

For Mary E. McDougall
1910-1995

Contents

Preface

Ted Rogers has not suffered from a lack of attention since he began to build his telecommunications empire in 1960. Especially after 1980, when Rogers Cablesystems went public and started to grow at an astonishing rate, business journalists have turned their gaze towards Rogers every time he acquired another company or reorganized the ones he already owned.

Over this period, he has maintained remarkable control over his image in the media. Almost without effort, he has persuaded one journalist after another that he's really just a nice guy with a passion for bringing new technology to the Canadian consumer. As a result, articles on Ted Rogers bear a striking similarity, whether they were written in 1980 or 1995.

My intention has been to cast a more irreverent eye upon the life of one of Canada's most fast-moving entrepreneurs. I have sought to illuminate his achievements in the context of Canada's business and communications communities, and within the cultural environment from the 1960s to the present. In the process, I have also tried to help the reader to understand the technologies of telecommunications and the regulations with which Canada's government tries to control it.

The work of other journalists, including David Olive of the *Report on Business* magazine, John Partridge of the *Globe and Mail,* and freelance writers Ross Fisher and Daniel Stoffman, has greatly contributed to the success of this endeavour. Any shortcomings are my own.

I wish to thank Peggy Sinclair for her meticulous research, which has continued through the final stages of this project; Meg Taylor, for her astute and sensitive editing, always delivered with welcome good humour; and Joshua Samuel, who has urged me to share the courage of his publishing convictions.

CHAPTER 1

The Present

Canada's $1-billion Man

The human race is faced with a cruel choice: work or daytime television.

ANONYMOUS

Ted Rogers arrived early on a Thursday in March 1995 at the CRTC hearings in Hull. Armed with a briefcase full of notes, he had answered the call to yet another session in the endless stream of inquiries conducted by Canada's communications regulators. For the third time in six months, he entered the fluorescent-lit hearing room to deliver his educated opinions about the business of packaging, distributing and enhancing information for Canadian consumers. He was also prepared to be photographed, interviewed, questioned and criticized about one aspect or another of his own role in the competitive and lucrative business.

Wearing a grey suit, white shirt and dark blue tie bearing a pattern of little emblems from the Royal Canadian Yacht Club, the sixty-one-year-old president and CEO of Rogers Communications Inc. looked at first glance like the quintessential white man of Canada's business establishment: fair haired, blue eyed, tall, fit and slightly geeky, with the rosy cheeks that upper-crust men develop after years of consuming rare roast beef and repressing long-forgotten passions. He was known for his iron will, icy sense of humour and devotion to Money with a capital M. Like other bluebloods of Canadian business, Rogers looked as if he knew which fork to use when they served the trifle at the Albany Club. Even if he had a couple of glasses of wine at a hoedown with the Eatons and the Bassetts in Forest Hill, you knew he wouldn't suddenly belt back a mouthful of Beaujolais, flip out and start playing air guitar when the kids cranked up the volume of their Aerosmith CD. Rogers hasn't flipped anything more than a fried egg since he was a boarder at Upper Canada College fifty years ago, and he wasn't about to abandon the conventions of behaviour and decorum that have helped him win the support of bankers, investors and corporate executives as he builds his company into one of the most successful owner-managed communications empires in the world. When you owe the bank $2 billion, as Rogers did this morning, you keep your shirt buttoned, your wingtips shined and your eccentricities out of sight.

But looks can be deceiving. Within the chest of every successful man beats the heart of a little boy, and Ted Rogers hasn't quite finished growing up. Despite the accumulation of fifty years of experience, this is the same guy who entertained his friends after curfew at Upper Canada by stringing up an antenna outside his dormitory window so they could watch programs from Buffalo on a TV set hidden in Rogers's closet. This was the guy who formed a university political party and named it after a popular brand of beer, and who followed John Diefenbaker into the bathroom when the Chief retired for a leak in a futile effort to avoid young Rogers's questions. As recently as fifteen years ago, Rogers had tried to wake up the fossils who owned a company in B.C. that he wanted to buy called Premier Cablevision by parading into their boardroom with a three-piece Hawaiian band and four hula dancers. Like Bob Dylan, who's only a few years younger and who went electric about the same time as Rogers went cable, Rogers cannot watch television for more than five minutes. If he ever watches a movie, it is likely to star Sylvester Stallone or Clint Eastwood. And even though the technologies on which his company has succeeded could make the printed word irrelevant, he still reads books. In fact, if he hadn't read about cellular telephony while taking a forced vacation in Australia with his wife, Rogers might never have started Cantel, which now contributes more than a third of his company's cash flow.

Cantel was only one of several things on his mind today as Rogers entered the CRTC hearing room. As usual, his brain cells were in overdrive, bouncing from one idea to the next like India-rubber balls in a squash court. But nothing about the man betrayed the complexity or number of his thoughts, except perhaps his readiness to smile. In no particular order, Rogers was also thinking today about: his competitors, his finances, his family, his company's image, satellites, telephones, interactive television and AT&T's announcement that it would encircle the African continent with 15,000 miles of fibre-optic cable. He liked that one. The sheer audacity of the plan appealed to him, even if it presented one more competitive threat to one aspect or another of his business.

If a man distinguishes himself by his competitors and critics, no one in Canada can measure up to Rogers. While most of us try to be good little Canadians who avoid calling each other names in public, Rogers has gone head to head in the public arena with politicians, regulators, journalists, academics, vested interests, entrenched ideas and bloated corporate bullies. They've called him a greedy capitalist, a Detroit knee-capper, a Watergate plumber and a devil with rabbit-ear horns and a coaxial tail. In the face of barbs from journalists and university pundits, who hold little influence over his own corporate future, Rogers usually remains silent. But at opportune moments, he will respond in kind, especially if he has a chance to hobble his competitors and grab a headline with a well-placed zinger. He once called the CRTC "the evil empire" and referred to his fellow private broadcasters, who continue to fight for a levy on cable companies to pay for local TV service, as a bunch of pickpockets. And if he's a Watergate plumber, he said recently, the dirty tricks of Canada's phone companies make them the corporate equivalents of Richard Nixon.

In his lexicon of rejoinders, though, Rogers holds a much more potent tool: success. Like a hockey player responding in the dying moments of a game to the taunts of an opponent by pointing at the scoreboard, Rogers merely has to draw attention to the spectacular financial growth of his empire to put his critics in their place. As a well-bred gentleman, he seldom resorts to such bluster. But his shareholders wouldn't mind. In the last twelve years, the value of their shares has increased from $7 to $140.

Starting in the 1960s, Rogers has expanded his company from a penny-ante radio station and supplier of elevator music into a billion-dollar enterprise by wrestling and cajoling for control with far better-financed and longer-established companies than his. In the fragmented Toronto cable market where he began, he battled for customers against Gulf & Western and CBS. He continued through the next two decades buying cable operations in Canada and the United States, backed by his friends the Bassetts and later by more adventurous high-fliers like the Belzbergs and Michael Milken. In the 1980s, with his company generating less than $500 million in revenues, he continued to beat far larger companies for the right to compete for long-distance telephone customers. And when the board of his own company scoffed in 1983 at his proposal to start a cellular telephone operation, he took their advice like a good soldier, then went around the corner and started Cantel.

By 1995, Rogers had become the Honest Ed of information. You want radio? We got radio. You want TV? We got TV. Wanna talk on the phone? Here, have one of mine. How about those Sabona copper bracelets? You can pick one up cheap from the Home Shopping Network. I own that, too. Long distance? No problem. And while you're at it, have a copy of the *Medical Post*. "There isn't anything quite like Rogers in the United States or anywhere else that I know of," said communications expert Eli Noam in 1993. "That's where the future is going to be: multiservice companies that do a bit of every-thing. Furthermore," added Noam, director of the Center for Telecommunications and Information Studies at Columbia University in New York, "if a company doesn't do it first, it won't be long before one of its competitors moves onto its turf."

For Rogers, each of his operations is "a different chapter in the same book. They're all communications," he says. "They are highly capital-intensive, have a high degree of technological change and a high degree of regulation."

To keep the capital flowing, Rogers has borrowed as fast and as much as he can, and probably more often than some of his supporters would prefer. Along the way he has mortgaged his house and hocked the family jewels and, on more than one occasion, stared bankruptcy in the face and felt the tingle in his nerve endings that comes when you owe a week's salary to your employees and there's no money in the account to pay them. Even today as he slipped into his seat at the front of the hearing room, his company's debt-equity ratio was an astounding 4:1, and the interest-rate meter was ticking at a rate of $1 million a day.

With his four children now in their twenties, Rogers didn't worry so much about shielding them from the debilitating glare of publicity. But he was somewhat miffed by an item in *Frank* magazine reporting that one of his daughters had tagged along with a gaggle of reporters from *Maclean's* magazine to interview the prime minister. So what if he *had* pulled a few strings to include her in the entourage? He owned the company, didn't he?

Still, he hardly needed another blow to the image of Rogers Communications. The company had taken a public-relations sucker punch when Rogers announced in late 1994 that customers would automatically receive seven additional channels, at an additional cost of $2.65 a month, unless they returned a card to inform the company they didn't want the new service. Rogers and most other cable companies had used the same approach, called negative-option billing, in the past. But this time it backfired. Rogers was portrayed as an underhanded money-grubber. Journalists immediately took the company to task for preying on unsuspecting citizens. What about the TV viewers whose mail doesn't reach them safely? they whined. Or the cable subscribers who throw out their response cards from Rogers along with all the other junk mail that pours through their mail slot while they're trying to concentrate on *Married . . . With Children*? Heck, what about the millions of TV viewers who can't concentrate long enough to read anything longer than a hockey score? They could end up paying for a deluxe visual pizza that they didn't order.

Judging from the uproar that ensued, Canadians cared far more passionately about the billing procedures of their cable company than they did about the national debt, Quebec's separation, Bosnia, the Islamic jihad, Somali warlords and the education of their children

6

combined. They ranted and raved and wrote letters to the editor in unprecedented numbers, most of them taking direct aim at the king of cable, Ted Rogers. In January 1995, more than 4,500 Canadians turned their attention away from *The Simpsons* long enough to write to the CRTC. Another 4,400 called the CRTC's offices by phone, some using long-distance services provided by Rogers and his associates through Unitel.

Rogers backtracked furiously, but it wasn't enough to stop politicians in Ontario, Alberta and B.C. from rearing up on their hind legs and barking in the name of truth, justice and their chances in the next election. In B.C., where the NDP government was embroiled in scandals involving sex, lies and patronage, politicians passed a law specifically outlawing negative billing, while calling Rogers an arrogant bully as they stood up for the little guy and hoped the little guy would forget their pecadilloes and remember all they'd done for him when the votes were counted.

Rogers's image had already been pummeled earlier that year when the company paid $350,000 for a condominium in a Vancouver building wired for cable TV by Rogers's arch competitor, B.C. Tel. The phone company was charging residents 40 per cent less for access to more cable channels than Rogers provided. But Rogers claimed the phone company had thwarted the public's choice for cable service and took his rivals to court, accusing them of oppressive business tactics. For the locals, this was better than watching a couple of naked amazons in a mud-wrestling contest. Journalists who didn't know the difference between a preferred share and a debenture accused Rogers of using his Toronto-honed muscle to stomp on the homegrown phone company. It became even more absurd in their eyes for Rogers to cry foul when, in February 1994, he launched a bid for Canada's most venerable magazine publishing company, Maclean Hunter, starting with an initial bid of $2.8 billion. How could you compete unfairly against someone who had $2.8 billion in his pocket? The public just didn't get it.

But the phone companies did. So did snipers like the Friends of Canadian Broadcasting, who regarded Rogers as a meat-eating psychopath, belching and farting his way through a side of beef at a love-in for vegetarians. Even though everyone in the communications industry knows that Rogers is still a pip-squeak compared to the phone companies, there were no Friends of Rogers Cable to set the public straight with a high-powered propaganda campaign. Meanwhile, the phone companies were successfully soft-selling the public with touchy-feely ad campaigns in which new mothers thanked their lucky stars for the nice telephone operators who let their week-old snookums exchange gurgles man-to-man with old gramps in his flannel shirt, somewhere in the bucolic potato fields of the mind. While Rogers blundered through the labyrinth of public opinion like a punch-drunk barbarian on steroids, the phone companies leaped and twirled and pirouetted like clean-living nimble ballerinas.

A similar hubbub was gathering in the United States. Unlike many of their Canadian counterparts, reigning politicians in that country regard barriers to competition as a communist plot. So U.S. phone companies have had far less trouble convincing their regulators of the need for competition in the cable industry. In the United States, competition keeps you strong and healthy and off welfare. U.S. consumers waited with open arms for someone to dance them away from their cable suppliers. In a survey reported in early 1995 by the *New York Times*, 55 per cent of cable TV households said they're dissatisfied with their TV programming and with the way it's provided by their cable companies. Given competitive prices, the survey indicated, these consumers would readily abandon cable in favour of phone or satellite companies. Like their counterparts in Canada, U.S. regional Bell companies all plan to offer video programming in the future, and almost 1 million U.S. households have already purchased direct broadcast satellite (DBS) dishes from companies like General Motors, Hughes and Hubbard Broadcasting. Canadian consumers, apparently, were no different. In a *Toronto Star* poll in March 1995, 75 per cent of respondents said they'd rather subscribe to cable services from Bell Canada than from Rogers.

And if that weren't enough to give Ted Rogers a case of the screaming abdabs, a woman in Coquitlam, B.C., was blaming her husband's death from brain cancer on his use of a cellular telephone. Where would it all end?

Not today, it seemed, and certainly not with the CRTC, which blew hot and cold when it came to supporting Rogers. Today, for example, Rogers's only apparent ally in these continual rounds of posturing and bullying was the chairman himself, Keith Spicer, a self-promoting former Trudeau-maniac who could make or break a humble subject within Canada's regulated communications empire with a nod of his big-domed head. But Rogers knew that Spicer was a fickle friend. He might love you today, but tomorrow was another story.

In the navel-gazing world of telecommunications, Spicer claims he answers to three masters: the creative community – shuckers and jivers who put their dubious talents together to dream up shows like *The Shirley Show, Wok With Yan* and *The Littlest Hobo* and then shove them down our collective throats; the distributors – companies like Bell Canada, Rogers Communications and, more recently, Power Corp., AT&T, General Motors, Ontario Hydro and dozens of others that have the wires, switches, tubes, conduits, satellites and bundles of money necessary to build and maintain the channels through which we receive our incomparable dog's breakfast of TV programming, fax messages and telephone calls; and the public, the undefined, inarticulate mass of humanity who irradiate their feeble brains in the chemical flicker of their TV sets or jabber non-stop from the driver's seat of a Taurus on their cellular phones. Given this collection of mismatched constituents, it's no wonder that the allegiances of Chairman Keith bound from one bailiwick to the next without rhyme or reason.

Today, Rogers wanted to put forward his view of the future of the information highway. That was the point of these hearings. Canada's government had commanded Spicer to spend a few weeks gathering information about the world of information. He then had forty-five days or so to condense this information into a definitive report that the government could use for policies to encourage growth, preserve culture, ensure competition, provide high-quality service, maintain employment, guarantee consumer choice, raise Elvis from the dead and ensure that the Montreal Canadiens won the Stanley Cup in 1997.

Rogers was as well placed as anyone in Canada to illuminate the future of the information industry. Ever since he'd purchased his first radio station, information in one form or another had provided the basis of his livelihood. Over the years his company had gathered information, packaged information, transmitted information and provided the technology for communicating information. Whether anyone really needed the information was not his concern. Rogers simply knew that people all over the world liked to talk, watch and listen until they fell off their chairs in a stupor. If he could provide them with the channels, the routes and even the vehicles themselves, people would line up to tootle down the information highway. All he had to do was stand back and collect the tolls.

Now, after thirty-five years, his company, Rogers Communications Inc., with revenues of $2.25 billion and an unaudited operating income of $722.4 million, has its corporate fingers in four distinct pies:

Wireless communications. Just as you might think, this describes the segment of RCI that provides services that don't require wires. What these services do require are the transmission facilities to bounce signals from one place to the next; gobs of money to pay for them; and lots of people talking non-stop at $1 a minute to provide the necessary gobs of money.

It may seem like a no-brainer today to put a phone in a car with a man by himself in a traffic jam. What else does he have to do except listen to self-help tapes on "How to Make More Money" or wallow in nostalgia as he listens to the Eagles play "Take It Easy" for the seven-hundredth time this week? But when Rogers did it, no one really knew if people would pay $1,000 a year or more to stay in their cars instead of running to a phone booth every time they had an idea.

It didn't take long for Rogers and his financial backers to see that they had a hit on their hands. Today, nearly 800,000 Canadians subscribe to Cantel, which is the only company licensed to provide subscribers with both cellular and paging services nationwide. Rogers Cantel Mobile Communications Inc. is Canada's largest wireless telephone company and has national licences to operate a digital cordless telephone service and an air-to-ground telephone service, both in developmental stage. Total revenue from wireless amounted to $750.4 million in 1994, and wireless operations contributed $290 million – more than 38 per cent – to the total operating income of Rogers Communications.

Visitors to the wireless world of Rogers pay a basic admission fee, and then they pay for each ride they take on the wireless Ferris wheel. This includes airtime usage, long-distance charges, optional services, system-access fees and roaming charges.

Of the company's 800,000 subscribers, more than 220,000 signed up in 1994. Many of them joined when Cantel started offering cellular phone services to customers beyond the conventional business community. Through 3,000 consumer outlets across Canada, Cantel aimed its AMIGO service at consumers who may have nothing to sell, no urgent information to convey – who may, in fact, have nothing at all to say – but who might use their cellular phone on weekends or evenings, when demand is low anyway. Some of these rookie subscribers signed up before they really knew what they were doing, and they were surprised when they had to pay for the service. "The hell with this," they said, when they found they had to pay $47 just so they could shoot the breeze with Aunt Barb on their way to Ikea on a Saturday afternoon. As a result, Cantel soon started receiving calls to disconnect the cellular service – more than 10,000 of them by the end of 1994.

The subscribers who stayed with Cantel, however, seemed to be a garrulous bunch. Every month they jabbered and yakked for 158 minutes apiece – over 2.5 hours – in 1994. This was five more minutes than the average in 1993. But because the company promoted its services heavily among people with less need for a cellular phone, and offered unlimited weekend and evening usage, Rogers earned a monthly average of only $77 per subscriber in 1994, down from $83 in 1993.

To offset revenue declines, Cantel dreamed up a number of new services, such as voice-dialling and call completion after directory assistance. The company also installed automated voice response systems for basic customer inquiries. And the future could hold even more promise for wireless. It's now possible to receive information of all types, from a rerun of *The Gong Show* to a spreadsheet from a branch plant in Tokyo to a phone call from your mother, in a variety of configurations over a variety of media. You might send and receive this information over a phone line using your personal computer, or over a TV cable using your television set, or over a phone line linked to your TV, or over a cable linked to your fax machine. Some people, however, think that wireless technology will play a major role in disseminating all this stuff throughout the civilized world.

In the United States, the federal government recently auctioned ninety-nine licences to companies that want to offer wireless personal communications services. In return, the government received a total of $7 billion. The top three bidders were Wirelessco, a coalition led by Sprint telephone system that includes three cable companies – Comcast, Cox and TCI; AT&T; and PCS Primeco, a consortium made up of NYNEX, Bell Atlantic, Air Touch and US West. Noting that the Sprint coalition includes three big cable companies, Reed Hundt, Federal Communications Commission chair, said: "The auctions just created the single largest wireless company in the world, and it's the cable television industry. This is the place where actual convergence between telephone, cable and long distance is taking place."

Later, Sprint and its three cable-TV partners announced they would spend $4.4 billion to offer nationwide wireless, local and long-distance telephone services that could be combined on one bill.

Cable TV. After picking up 675,000 new subscribers in 1994 through the acquisitions of Maclean Hunter and West Coast Cablevision Ltd. in Burnaby, B.C., nearly 2.6 million subscribers, along with their hyperactive kids, now rely on Rogers Cablesystems to keep them plugged in to quiz shows, soap operas, movies and sports broadcasts, twenty-four hours a day, 365 days a week. Rogers customers account for nearly 33 per cent of cable TV subscribers in Canada. Needless to say, Rogers is the country's largest cable-TV company, focused on major urban areas in Ontario and B.C.

Cable companies swap subscribers the way the Blue Jays trade pitchers. In 1994, for example, Rogers sold its cable operations in Calgary and Victoria to Shaw Communications. It also acquired cable operations in Thunder Bay and Sault Ste. Marie, Ontario, from Maclean Hunter, then turned around and sold them again to Shaw. In return, Shaw coughed up its cable holdings in southwestern Ontario and Metro Toronto. Shaw emerged from the deal with an additional 149,000 subscribers, for which it paid Rogers $201 million, about $1,350 per subscriber.

In addition to cable service, Rogers also provides pay-TV and pay-per-view services and converter rentals. Hotels hire a subsidiary called Rogers Network Services to pump movies into the pay-TV in guest rooms. The subsidiary also provides point-to-point telecommunications. Although they generate only a small percentage of Rogers Cablesystems' total revenue, a chain of 110 video rental stores with the surprising name of Rogers Video gives the company access to an extensive video library.

In 1994, cable accounted for 43.2 per cent of Rogers total operating income, equivalent to $376.3 million. From cellular and cable combined, Rogers earned more than 90 per cent of its operating revenues.

Even after two increases in its monthly rate in 1994, Rogers retained more than 93 per cent of its subscribers. And revenues per cable subscriber increased by 9.1 per cent. As the company imposes further rate increases, some subscribers will unplug their cables and buy satellite dishes. If the phone companies enter the market, many more Rogers customers will turn to Ma Bell. Some may simply turn off the TV and never turn it on again (but no one will hear about those two people).

Meanwhile, the company's costs are rising. In 1994, the cable company's total expenses increased by 7.1 per cent. Among other things, this reflected an increase in the fees paid to suppliers of specialty television programming. Capital expenditures on upgrading the cable network, increasing channel capacity and expanding services also rose by 87 per cent to $247 million. And Rogers spent heavily to expand its pay-per-view channel capacity to twenty channels from four, allowing the company to compete with direct broadcast satellites by late 1995.

For years, the federal government was firm in prohibiting U.S. satellite providers from selling subscriptions in Canada. But as Prohibition, the tobacco industry and the Paul Bernardo media circus have shown, the market will prevail. It began with some enterprising Canadians who just up and drove across the U.S. border to Blaine, Washington, or Buffalo, New York, purchased a DirecTV receiver, installed it in their Canadian home and arranged to receive bills through a U.S. address. Meanwhile, Expressvu Inc., owned by a consortium of telephone companies and broadcasters, and Power DirecTV, a Canadian affiliate of DirecTV in the United States in partnership with Power Corp., prepared to offer satellite service in Canada as soon as they received the go-ahead from the CRTC. That ruling came through on July 7, 1995. Within months, both companies would be operating in Canada.

Multimedia. Rogers set up Rogers Multi-Media Inc. in late 1994 primarily to accommodate the publishing interests of Maclean Hunter. Multi-Media includes TV and radio broadcasting, periodicals for Canadian consumer and business readers and a 62 per cent interest in the Toronto Sun Publishing Corporation. Multi-Media broadcasting operates twenty radio stations (eleven AM and nine FM outlets) and three TV stations, with minority interests in other specialty programming services like YTV and the Home Shopping Network. The publishing arm produces seven consumer magazines, including

Chatelaine and *Maclean's*, and thirty-five specialized business direc-
tories and periodicals, such as *Monday Report on Retailing* and the
Medical Post. The newspaper operations publish ten daily papers,
including the *Financial Post* and the *Toronto Sun*, sixty-four weeklies
and shoppers, and a variety of specialty magazines. Compared to
cable and wireless operations, Multi-Media contributed a paltry 8.9
per cent to Rogers's total operating income, from total revenues of
$680 million.

Long distance. As of July 1995, Rogers owned 29.5 per cent of Unitel
Communications Holdings Inc., providing public switched and private-line voice and data services through a national, facilities-based
digital network. At the time, other shareholders were Canadian
Pacific Ltd. (48 per cent) and AT&T (22.5 per cent), although CP was
seeking a buyer for its shares.

Unitel holds a 6 per cent share of the public switched long-
distance market and a 22 per cent share of the private-line voice and
data market, but has never made money. In 1994, Rogers's share of
Unitel's losses amounted to $76.9 million, up from a loss of $66.4 million in 1993. The 1994 loss amounted to about 10 per cent of RCI's
operating income. In addition, Rogers contributed $103 million to
enhancements and write-offs at Unitel. (In August 1995, RCI wrote off
its investment in Unitel.)

Despite the extent of his holdings, Rogers has sensed for years that
his company might soon become roadkill on the information highway. Like a deer caught in the lights of an oncoming Mack diesel, he
could see the imminent threat of satellites, phone companies, home
computer providers, even utility companies. The U.S. power industry,
for example, announced in March 1995 that it would build the information superhighway. In fact, if the utilities had their way, kids would

loll around the kitchen watching *Mighty Morphin Power Rangers* on the pop-up toaster; Dad could sit in the backyard sipping on a beer and watching the Blue Jays on the gas barbecue; and Mom could watch *Home Improvements* on a telephone with a six-inch screen.

And every day another threat appeared on Rogers's threshold. People had started making long-distance calls on the Internet, using modems supplied by Motorola. Thieves were scanning the airwaves for phone numbers and security codes, then stealing as much as $1 billion a year from the North American cellular phone industry.

In early 1995, MCI Communications and the U.S. Public Broadcasting Service announced that they would provide television programming via modem to home computers. DirecTV satellite service was offering 144 channels to subscribers in the United States, Central and South America, and more than 80 per cent of U.S. satellite subscribers said they were satisfied with the service and would not switch back to cable even if the cable industry handed them a cheque to cover the money they'd spent on their satellite receivers and other paraphernalia. AT&T, an ally of Rogers in Unitel, could send video signals through existing copper-wire networks. Other global behemoths, including NTT in Japan, British Telecom in the U.K. and Deutsche Telekom in Germany, were striding through the information universe stomping on companies like Rogers and Bell Canada as if they were small bugs. Of the top twenty telecommunications companies in the world, eleven are based in the United States, which Rogers could see on a clear day from his office. The Friends of Canadian Broadcasting might pull their beards and shuffle their Birkenstocks and complain about Rogers's dominance of the Canadian cable industry. But compared to the cast of leading characters on the worldwide communications stage, Rogers was still nothing more than a bit player mumbling to himself in a crowd scene.

Clearly, Rogers had a fight on his hands, and if Canada's regulators didn't know it, he would give them a ringside seat. Flanked by his trusted lieutenants, Colin Watson, president and CEO of Rogers Cablesystems, and Phil Lind, vice-chairman, who had served the company for a combined total of fifty years, Rogers did his best to explain why Canada's cable industry needed the CRTC's protection from the rapacious technological claws of Canadian phone companies. The phone companies were panting hungrily for a chance to compete with Rogers in the market for cable services. They had the technology. They had the public support. They had the money. And they had the ardent desire. All that stood in their way was Spicer's intransigence. Earlier that week, the phone companies had stepped up to the podium to give Spicer their side of the story. They explained why cable companies in general and Rogers in particular should not enjoy another minute of a monopolized cable market. In the process, they compared Rogers to a Watergate burglar and their long-distance competitors, of which Rogers was one, to regulatory junkies addicted to red tape, who would steal the hubcaps off your car as you rolled down the information highway.

Now Rogers wanted to set the record straight. Contrary to public perception, said the man who had just bought a $3-billion company, Rogers Communications was not rolling in dough. Unlike Canada's six largest phone companies, through which cash flowed like scented water at the rate of $2 billion a year, poor little Rogers and its five most prosperous associates in the cable industry saw a mere trickle of $4 million. The telephone companies have retained earnings of $4.2 billion, compared with the cable companies' $95 million, he added. "The cable industry financially is not survivable in an open street brawl starting tomorrow," said Rogers.

As if that weren't enough to set the alarms off in Spicer's head,

Rogers told the chairman that the phone companies didn't play fair. They use the money they collect from local services to subsidize their long-distance service, he said. They price their competitive services below cost. Gee willickers, he added, sometimes the phone companies won't even let a Rogers guy climb their poles to hook up the cable doo-hickeys. What's a guy to do, anyway?

CHAPTER 2

Corporate History

Surfin' Safari

*New communications technologies arrive in waves, and
Rogers is the best surfer in the business.*

DANIEL STOFFMAN

Report on Business magazine, September 1989

Ted Rogers stormed into the information age in a typically Canadian
way – slowly, methodically, at great expense, without fanfare and without much help from Canada. Most other Canadians took almost
twenty years to realize that the information age had arrived. In 1962,
a quirky Toronto professor named Marshall McLuhan said, "The
medium is the message." Two decades later, we discovered one of our
own boys smack-dab in the middle of it. And he hadn't slithered off to
Los Angeles or London; he was still right here in Canada. Over the
years, Canada had deprived him of money, tied him up in red tape,
dropped regulatory bowling balls on his toes, called him names and
sneered at his accomplishments, and still he'd continued his relentless march towards the new age of communications.

Even now, Canadians see Rogers less as a visionary and more as a threat to their well-ordered world. "Rogers may be a visionary, but he's a business visionary, not a true communications visionary," observes a much-quoted visionary expert and law professor at the University of Toronto named Hudson Janisch. Like Janisch, most Canadians regard business as an arena for blockheads, and unbridled competition as an affront to their dignity. In Canada, unless you're Paul Anka, Eric Lindros, Silken Laumann, or Northern Dancer, the single-minded pursuit of a dream is considered a compulsive addiction. But even the wet blanket of the Canadian psyche couldn't repress Rogers's drive. It just slowed him down a bit. In the eighteen years between 1960, when he took his first plunge into the world of commercial information, and 1978, when he acquired a public company called Canadian Cablesystems Ltd., Rogers learned how to navigate Canada's financial and regulatory pathways until he emerged with the largest cable TV operation in the country. Then he could really go to work.

Like Canadian rock stars and movie directors, Rogers also had to prove himself in the United States before anyone would pay attention to him in Canada. In fact, if it weren't for the regulations that Canada belatedly imposed on the business of culture, most of Rogers Communications would likely be owned by Americans. Except for a few friends and Canadian punters, only U.S. investors would risk their money on Rogers's vision before it became a sure thing.

Rogers staked his first claim, in 1960, on a minuscule plot of the information bonanza called CHFI, an FM radio station in Toronto. Canada hardly noticed. To the ordinary tax-paying, church-going Canadian of the 1960s, the ramparts of the nation's information business were anchored by such monolithic structures as the CBC, Bell Telephone and Maclean Hunter. These were venerable, law-abiding organizations led by steely-eyed, close-mouthed old men in shiny black shoes who sat behind big desks in wood-panelled offices on the top floor of their concrete buildings dictating important letters to their purse-lipped, fastidious secretaries, who were all named Miss Miller and wore their hair in a bun. Under such stewardship, these organizations hired thousands of Canadians and employed them for life, keeping them informed with a monthly employee newsletter about company bowling teams and new members of the twenty-five-year club, rewarding their loyalty with a regular paycheque and never making them work on Sundays unless it was absolutely necessary. Employees' sons and daughters followed their parents into the corporate womb, never to be heard from until they emerged forty years later with a gold watch or an engraved pen set from Birks. School teachers and bus drivers could invest a few dollars in a share of Maclean Hunter or Bell Telephone and sleep soundly for decades, reassured by the dubious conviction that they'd still have a few dollars when they woke up again.

In the shadow of these ramparts, a few grub-stakers pitched their tents and began hawking their own variations of contemporary information. With names like Moffatt, Griffiths, Greenberg, Bassett, and Sobol, they built buildings, constructed studios, erected antennae and tried to persuade listeners and viewers within range to tune in. Initially, their radio and TV stations operated for less than fifteen hours a day. At midnight, after listening to "God Save the Queen" and the politically incorrect version of "O Canada," you could hum along to the monotone test signal or spend the next eight hours staring at your Philco TV screen mesmerized by a black-and-white test pattern that resembled the face-off circle at centre ice in Maple Leaf Gardens.

Even at this rinky-dink stage of development, it was expensive to buy the equipment and hire the people to run a competent broadcasting operation, and most of these upstarts used their parents' money to help them get going. Rogers started comparatively small and even farther beyond the ramparts than his contemporaries. His family's war chest wasn't as full as the Bassetts' or the Eatons'. Nor could he model himself on a living, breathing father like John Bassett, publisher of the Toronto *Telegram*, or John Eaton, Mr. Department Store. His own father, Edward Rogers Sr., a radio pioneer who developed the world's first plug-in radio and founded the world's first alternating-current radio station, had died more than twenty years earlier, at the age of thirty-eight, with no insurance and few provisions for the family he left behind. Rogers's mother sold the radio station, CFRB, to Standard Broadcasting, and Rogers, who was five years old, was left to find inspiration in a memory and a handful of vacuum tubes, fuelled with a sense of injustice that such a talented man as his father should die so young and leave his family so bereft. Though he hadn't left much money behind, Rogers's father did bequeath to his son a passion for tinkering with tubes and transmitters and speakers. This talent for stringing a length of wire out the window, plugging another wire into an electrical socket and making sounds and pictures emerge from a gizmo gave Rogers a perspective on the world of communications that most other investors didn't have. "I didn't get into broadcasting out of any smarts," he said later. "I was emotionally attracted to it because of my father."

As Rogers prepared to take his first steps into the business world, in 1960, he already seemed to know something that his contemporaries didn't. While most would-be communications tycoons put their money into the conventional AM radio and TV set-ups of the day, Rogers paid $85,000 in inherited and borrowed money for a 940-watt FM radio station in Toronto whose listeners were restricted to a couple of ham-hackers from the University of Toronto's Faculty of Engineering and fourteen old geezers in their downtown Toronto basements reading *Popular Mechanics* and tinkering with their radio sets. "They were like the members of an exclusive club," Rogers observed. "They thought they had something that was extraordinary and unique. They didn't think they were better than other people, just more adventuresome and innovative." Rogers was twenty-six years old.

At the time, only 5 per cent of the population in Canada even owned an FM radio. Restricted to such an exclusive market, Rogers's station, CHFI-FM, generated annual revenues in 1960 of just $82,000. What kind of lunatic would pay good money for a station that no one could hear even if he wanted to? "Hell, it was broadband," Rogers explained. That enabled FM radio to deliver sound in stereo and high fidelity, enhancing the pleasure that listeners derived from their favorite ditties recorded by Mitch Miller, Mantovani, Bert Kempfert and Herb Alpert and, later, by Steppenwolf, Neil Young, the Bronskie Beat and Frankie Goes to Hollywood. "How in the world could it not win?" Rogers added modestly. "The trouble is, no one had an FM radio."

Some people might regard that as a major obstacle, like owning a hockey team in a world without ice. But if people didn't immediately flock to the local shopping mall to buy an FM radio, Rogers found other ways to make sure they could listen to his station. First, he arranged with Westinghouse to manufacture cheap FM radios for about $40 apiece. Then he either sold them at a discount or gave them away to potential listeners. Within two years, CHFI-FM was making enough money that Rogers could buy an AM station, supported by a loan from the Bank of Montreal. Nine years later, in 1971, Rogers would change the name of this AM station to CFTR (for Ted Rogers).

Even then, in 1962, there was no guarantee that his enterprises would grow. The battlefield of business is littered with the corpses and gasping hulks of companies financed with family money by privileged kids who assumed their own failure was genetically impossible. Given the fate of Creeds, People's Jewellers, CKO – All-News Radio, *Vista* magazine, Levy Industries and the Edsel, there was no reason to assume that Rogers, whose silver spoon was on the dainty side, would eventually run a communications empire that would employ more than 13,000 people and generate $2 billion a year in revenue by 1995; certainly not when, in 1960, he was still scratching his sandy-haired bugle over the theory of uses in his property-law class at Osgoode Hall.

Although he was driven by a different kind of intensity, Rogers maintained his friendships with the Bassetts and the Eatons. So when the opportunity came to join them in their fledgling Toronto TV station, CFTO, along with a popular Toronto commercial announcer named Joel Aldred, he didn't hesitate. It required a huge investment, provided initially by the Eaton family, to buy the transmitters, cameras and equipment they needed to start the station. But land was cheap on the northeastern fringes of Toronto, and television viewers certainly wanted another TV station. As suburbs arose in the farmers' fields and apple orchards that surrounded Toronto, every house on the block seemed to have an aerial, shaped like a gigantic potato masher, sticking up from the roof or bolted like an oil derrick to the side wall. These people had already seen John F. Kennedy, the handsome young airhead, turn the slack-jowled but far wiser vice-president of the United States, Richard Nixon, into a doddering fuss-budget with a five o'clock shadow in a televised debate in 1960. They appreciated the power of the medium, and they wanted more TV shows, in any shape or form, to distract them from the humdrum realities of life in the 1960s.

CFTO would provide those shows, although the Board of Broadcast

Governors, predecessor of the CRTC, would make sure that 55 per cent of the shows were Canadian. One of the first privately owned TV stations in Canada, CFTO (Channel 9) began broadcasting unforgettable fare like *Romper Room* and *Professor's Hideaway*, making TV viewers and CFTO's owners happy even while confirming the worst fears of Newton Minow, the Kennedy-appointed chairman of the U.S. Federal Communications Commission (FCC). At the age of thirty-five, Minow had told a gathering of broadcasters in Washington that television programming was a "vast wasteland – a procession of game shows, violence, audience participation shows, formula comedies about totally unbelievable families . . . blood and thunder . . . mayhem, violence, sadism, murder . . . private eyes, more violence, and cartoons . . . and, endlessly, commercials – many screaming, cajoling and offending." In short, TV delivered just what people wanted, especially in a city like Toronto, where nothing happened after dark, where taverns maintained separate entrances for women with escorts and closed at 11:30 p.m. even on New Year's Eve, and where, on a Sunday morning, you could fire a cannonball down Yonge Street without hitting a living soul until it reached Rochester on the other side of Lake Ontario. The vast wasteland on the tube was nothing when you considered the wasteland outside the front door. If viewers couldn't get their daily dose of mayhem and carnage and mindless idiocy on the street, they'd take it on the screen, whether it came from CFTO or more hard-core outlets of blood and thunder south of the border that delivered the Real McCoys and didn't dilute their programming with vapid Canadian substitutes.

For Rogers, programming of any complexion was a secondary issue, even when he shared ownership of his first TV station. His passion lay in the engineering, the doo-dads and dinguses that made it possible to record sound and pictures and hurl them through the air right into the living rooms of all the Mr. and Mrs. Potatoheads patiently sitting in their La-Z-Boys with their chips and Pepsi waiting to see what mischief Ricky and David would get up to today on *Ozzie and Harriet*. When people talked about the quality of television, Rogers thought about the clarity of the picture, not the intellectual content of *Razzle Dazzle*. So when he learned about cable and saw its potential for replacing the oversized rooftop spud mashers that defaced the suburban skyline with a technology that guaranteed a clear picture and more stations, he knew that people wouldn't hesitate to pay $4.50 a month for the service.

Until Rogers came along, Toronto TV viewers squinted at *Superman*, *Howdy Doody*, *Arthur Godfrey*, *December Bride*, *Our Miss Brooks*, *You're in the Army Now*, *The Honeymooners*, *Desilu Playhouse*, *Peter Gunn*, *The Lone Ranger* and *Milton Berle* through blizzards of electrical snow that hissed and crackled across their TV screens. If they weren't buried in fuzz, images on the screen pulsed and flickered, like the nerve endings of a drilled tooth, in response to interference from airplanes flying overhead on their way to Malton Airport or from Mom plugging in the Mix Master to whip up another angel-food cake. Sometimes the ghosting on the screen was so severe that it multiplied the number of characters by a factor of five, so you couldn't tell where the real Art Carney ended and his silhouette began. In some parts of Toronto, a TV set couldn't receive a signal from a U.S. station, no matter how many extra feet of tubing you added to the aerial. In futility, you could fiddle with your rabbit ears so they pointed west or east or north or south or all four at once, and you could even buy a revolving antenna that would spin around on your roof like a naval radar unit. But what everyone really needed was a link to a gigantic antenna somewhere in the middle of nowhere that would deliver clear reception come hell or high water, no matter how many

DC-3s flew over the rooftops or how many times Dad in his workshop decided to buzz through a few thousand feet of plywood with his power saw while you were trying to watch *American Bandstand*.

Typically, Rogers began thinking about his new venture while he was on a holiday. By 1966, Rogers had been married for three years and had hardly taken a day off. Working eighty hours a week, he had little time and even less interest in leisurely amusements. But his wife, Loretta, insisted that they take a vacation far from the intrusive demands of Rogers's business. "We went to Australia," Rogers recalled. Physically, he might have been with the kangaroos and koala bears halfway around the world, but his mind never left Canada. "While we were there, I read a little book on cable," he said. "When I got back, we applied for a licence."

With help from his friends the Bassetts and the Eatons, Rogers acquired a licence to provide cable TV services to parts of Toronto. At the time, two years before the Trudeau government set up the CRTC, cable was less regulated than pinball. To acquire the licence, Rogers had to present his case to the federal Department of Transport, which administered the process. Another company was applying for the same licence, but Rogers beat Maclean Hunter to the punch. The licence cost $25.

"You have to understand," said Colin Watson, now president and CEO of Rogers Cablesystems, "that in the mid-60s, cable was still a primitive business. There were maybe four or five companies cherry-picking areas of Toronto, like Hogg's Hollow, where U.S. signals couldn't be picked up."

With licence in hand, Rogers, at thirty-two, became a competitor not only of Maclean Hunter but of Gulf & Western, CBS and other much larger communications organizations. The competition didn't faze him. "I tend to work better when I'm up against a really big adversary," he said, "because I think then you really have to work and you really have to perform." Besides, despite the stature of his competitors, cable TV at the time was not a big business. It was a speculative technology at best, with all the cachet of a used car outlet. Most people weren't sure if the newfangled technology was all that it was cracked up to be, and they hesitated before committing themselves to another monthly expense. After all, TV still hadn't become an essential household appliance, like a telephone. You couldn't talk back to the tube or use it to order a chesterfield from Eaton's; you could only sit in front of it like a dummy, letting it talk to you. "Cable wasn't so much an industry as an act of faith," Watson recalled. Only Rogers and a few faceless corporations seemed to believe that cable would play a major role in the future of communications.

The licence gave Rogers the authority to seek 2 million feet of cable from Bell Canada, which had chosen to ignore the cable industry and leave it to the pipsqueaks and gamblers to test the technology. Bell Canada workers installed the cable for Rogers, stringing it from phone pole to phone pole throughout his licensed territory.

To pursue his ideas further, Rogers realized that he couldn't operate alone. In the aftermath of his father's death, he understood the limitations of a visionary who works in isolation. Until now, he had relied on close family and friends for advice. His wife was his right-hand person; his mother, Velma, "my pal and partner," provided inspiration and emotional guidance; his stepfather, John Graham, who had married his mother in 1941, when Rogers was seven, contributed his corporate wisdom and stability; and John Tory, whom Rogers had met as an articling student with Tory Tory DesLauriers & Binnington, contributed his own perspective and incisive advice. But Rogers needed a more formal management team to help him, and he began slowly to recruit its members.

Phil Lind was the first to join. An alumnus like Rogers of Upper Canada College and a brother of Zeta Psi fraternity – "the screaming Zates" – Lind had also attended Ridley College before enrolling at McGill University. At twenty-six, he had little knowledge of cable television, but then who did? In 1969, Rogers hired Lind to oversee the cable company's community programming.

In hiring Lind, Rogers wasn't simply adding a like-minded but junior WASP to his staff, someone who would agree with everything he said and step out of the way if the boss wanted to pursue another screwy idea. Although the cable business had been around for a few years, it still suffered from the dubious image associated with an unproven technology. Unable to attract top-notch experienced managers, Rogers turned to young, eager, ambitious and untested talents within his own circle of friends and acquaintances.

In addition to helping Rogers meet the CRTC's demands for community programming, Lind quickly assumed a strategic role within Rogers Cablesystems. During the 1970s, Rogers borrowed heavily against the cash flow that bankers had come to appreciate in the cable industry. People will always pay their cable bills rather than face a fuzzy TV screen. Meanwhile, Lind helped Rogers to expand his fledgling cable empire into Vancouver and the United States. "Rogers gave me a budget that he was prepared to spend up to and said 'Keep me informed,'" Lind recalled. "But he never second-guessed me."

Rogers had already plunged head first into an ocean of debt in 1970, when the CRTC first prohibited the cross-ownership of cable systems and TV stations. The regulation forced Rogers to sell his share of CFTO to the Bassetts and buy the share of Rogers Cable owned by the Bassetts and the Eatons. Through the Toronto Dominion Bank, Rogers and his wife eventually borrowed $2.5 million to conduct the buy-out, mortgaging their house in Forest Hill as part of the loan's security. Fortunately, debt didn't intimidate Rogers. "You don't deserve to be called an entrepreneur unless you've mortgaged your house to the business," he observed.

Throughout the 1970s, Rogers relied on debt financing as he aggressively expanded his cable empire, challenging companies much larger than his own and eventually getting his way. In 1973, he tried to persuade stockholders of Canadian Cablesystems Ltd. to sell their shares to him. At the time, Rogers ranked fifth among Canadian cable companies in terms of subscribers and revenues. CCL, which had been founded in 1920 as Famous Players Canadian Corporation, had 465,000 subscribers and was the second-largest cable company in Canada. CCL's management team, including a young whippersnapper and motorcycle enthusiast named Colin Watson, thought Rogers had a lot of nerve to try to buy a company so much bigger and better established than his own. Egged on by management, CCL's shareholders rebuffed Rogers's overtures and told him to fly his kite off someone else's antenna. "I shall return," Rogers vowed, and he headed west in a cloud of dust to conquer another contender for the cable crown.

In Vancouver, Premier Communications was run by George Fierheller, a contemporary of Rogers at the University of Toronto and a Sigma Chi fraternity member. The company provided clear TV reception to homeowners in the Fraser Valley, where TV signals bounced like pinballs off the majestic but pesky mountains that raised their jagged peaks to the sky like good scenery but made it almost impossible to receive a clear picture from Washington or Oregon to the south. Premier also operated cable systems in southern Ontario. In 1976, Rogers bought 20 per cent of the company and laid plans to buy even more.

Still a relative upstart in the industry, with about 200,000 subscribers, Rogers resumed his attack on CCL. From Edward and

Peter Bronfman, Rogers acquired 25.8 per cent of CCL in the name of his holding company, Rogers Telecommunications Ltd. (RTL). The Bronfmans, through Edper Investments Ltd., held another 24.21 per cent of the shares. Rogers and Edper agreed that one party would give the other first right of refusal to buy its shares.

CCL complained to Ottawa, seeking to have the arrangement over-ruled. But Rogers was prepared. He and his team of trusted advisers descended like the Flying Wallendas on the nation's capital, prepared to execute intellectual backflips to justify their approach to CCL. Rogers himself argued elliptically about the distinction between own-ership and control and pledged to expand CCL's channel capacity. Guided by John Graham, RTL's chairman and Rogers's stepfather, the team also argued that the acquisition of CCL would be good for the country. In case the CRTC didn't buy that argument, Rogers threw in $3 million in social benefits, including a satellite consortium for cable operators and a rationalization of Toronto's patchwork cable system. "I didn't think he had a chance," Watson said.

As it turned out, however, Rogers could have saved his breath. Eager to untangle the skein of conflicting territories that had turned the cable industry into a regulatory nightmare, and over the strenuous objections of CCL, the CRTC approved Rogers's $35-million purchase in January 1979. Meanwhile, Rogers exercised his option to buy out the Bronfmans for $18 a share, "less than $100 per subscriber," he said later. (A decade later, he would sell his U.S. cable systems for up to $3,000 per subscriber. In Canada, cable companies would value sub-scribers at about $1,000 apiece by 1989.) The acquisition vaulted Rogers into the lead in Canada's $235-million cable TV industry. Rogers Cable now served 17 per cent of Canada's cable subscribers and 37 per cent of the Toronto market. But it was hardly a corporate powerhouse. Generating revenues of $58 million, with a book value of about $100 million, dinky little Rogers still didn't pose much of a threat to the mighty oaks of the Canadian corporate forest like Bell and Maclean Hunter.

Over the same period, Rogers continued to acquire U.S. cable systems – in Syracuse, New York; Orange County, California; and Portland, Oregon. He also tried to set up a system in Minneapolis, but residents and local politicians resisted ownership of such a precious community resource as a cable TV network by a wild-eyed alien capitalist from some two-bit country where people put vinegar on their French fries and sing songs about a queen. Even after Rogers sent Phil Lind to assuage the natives, Minneapolis continued to turn its back on the suitor from Canada. But Rogers wouldn't take no for an answer. Within a year, he had changed the mind of Minneapolis, partly by promising to set up an institutional network of cable lines that would link businesses, schools, and community centres. When he made the promise, Rogers had every intention of fulfilling it. But when the recession hit in 1982 and forced him to cut spending drastically, Rogers had to back away from his promise, leaving local power brokers with their political shorts in a knot.

By then, Rogers had secured his hold on Canada's cable industry. In 1980, he bought the remaining 80 per cent of Premier Communications in Vancouver that he hadn't acquired four years earlier. Encouraged by Premier's president, George Fierheller, and prodded into action by marching bands and hula dancers parading through their boardroom, Premier's directors advised stockholders to accept Rogers's offer. In the process, Rogers also acquired a 12.5 per cent share of Vancouver's NHL hockey franchise, the Canucks. "I wish some of our other assets were performing as well," he quipped at the time, with an eye on the enormous debt that was piling up under his company's assets.

With his acquisition of CCL, Rogers had turned his holdings into a public company. In 1981, he renamed the operation Rogers Cablesystems Inc. (RCI), later to become Rogers Communications Inc. in 1986. With 1.2 million subscribers, the company was generating more than $70 million a year in revenues. But Rogers had set his sights even higher. Satellites like *Telstar* and *Anik* had been floating around in space for decades, and the customers of companies like CNCP Telecommunications had been dinging data off these disks for years. Rogers himself was one of them. The consumer in Oakville, Ontario, or Surrey, B.C., might never have mistaken his telephone for a TV set, even after one too many Red Caps. But to people in the telecommunications industry, such as Rogers, the distinction was far less apparent once you got beyond the instruments themselves. Microwave signals were microwave signals, whether they carried a conversation between Uncle Charlie in Nanaimo and his niece Mary in Toronto or the latest episode of *Seventy-Seven Sunset Strip* with Efrem Zimbalist Jr. and Edd "Kookie" Byrnes. The world of communications was changing, and the barriers between telephones, televisions, radios and telegraph transmission were disappearing. Already, Rogers was offering cable services that could link local bank branches and supermarkets in the U.S. at a lower cost than the competing telephone system. But Rogers's company still hadn't acquired the necessary critical mass to play a major role in the emerging new industry. He knew he was ahead of the pack. But it wouldn't take long for someone else to come along and sully his dream. "If some of these things had been more apparent then," he said, "others would have gone ahead and done them first."

In May 1981, Rogers pledged his family's radio stations and cable TV interests to secure a loan to buy 51 per cent of the ninth-largest U.S. cable TV company, UA-Columbia Cablevision Inc., in Westport,

Connecticut. Teamed with United Artists Theater Circuit Inc. of San Francisco, Rogers's bid of $90 a share beat by $10 a competing offer by Knight-Ridder Newspapers and Dow Jones. The deal, worth US$152 million, added 450,000 U.S. subscribers to Rogers's growing operation.

To raise money, Rogers sold the 49 per cent stake in Famous Players that he'd acquired with CCL, for cash and shares (including 600,000 shares of Williams Electronics, a manufacturer of pinball games and slot machines) worth more than $36 million. But Rogers still faced enormous obstacles to his objectives. He was pouring millions of dollars into his cable operations around the United States, installing cable and equipment and trying to attract subscribers. Some of these systems wouldn't break even for another six years, and Rogers knew he couldn't wait that long before someone else barged in to spoil his party. To alleviate his troubles, he approached his partner in San Francisco and proposed that they acquire the rest of Rogers's existing U.S. assets. United Artist Theater Circuit refused the offer. The company didn't mind climbing into bed with Rogers to consummate its deal for UA-Columbia, which would start to generate money almost immediately. But it had no desire to make room for all of Rogers's poor relations too.

By 1982, Rogers had become the largest cable operator in North America. At one point, Rogers's subscription list surpassed 2 million households, making it the largest cable company in the world. For the first time, Rogers Cablesystems ranked among the top 500 companies in Canada, with sales of $109 million and assets of $421 million. In twelve years, the revenues of Rogers's company had risen by 1,000 per cent. But to get there, he had assumed a burden of debt that had increased by fifty-two times in the previous five years, to more than $750 million.

Until interest costs exceeded the company's operating income, debt had been no problem. But in 1982 and 1983, another recession hit Rogers like a tornado. Interest rates rose to around 20 per cent, and the cost of Rogers's money, much of it borrowed at floating rates, rose along with them. For the first time since he'd scrambled to buy out the Bassetts and the Eatons in 1970, Rogers wasn't sure if his company would survive. In near panic, he began to unload assets as fast as he could. He even arranged with Bell Canada to postpone payment of the fees that he owed the phone company to keep his cables dangling from Bell's poles.

Rogers also raised about $300 million to pay down his bank debt through a high-flying investment dealer in Beverly Hills named Mike Milken of Drexel Burnham Lambert. Over the years, the much-maligned Milken would raise billions of dollars for upstarts in the communications industry, such as McCaw, Turner, and Viacom, which couldn't qualify for a conventional bank loan. A $300-million deal for this Canadian dude named Rogers was a moment's diversion: as Rogers and his senior vice-president for investments, Graham Savage, stood by his side, Milken raised the money in less than half an hour.

Back in Canada, the federal government had licensed six pay-TV networks to great fanfare in 1982. One network was national, three were regional and two were special interest. At the time, pay TV was touted as the next best thing since the BeeGees. Government and industry alike had witnessed the astonishing success of cable, and this would be even better. Great movies; no ads; right in the comfort of your living room: How could it lose? The government predicted that pay-TV would generate $600 million in revenues by 1987, of which $236.8 million would go to cable companies.

Unfortunately, what the government gives with one hand it takes with the other, and pay-TV was no exception. Under CRTC rules,

40 per cent of pay-TV's content had to be Canadian. Even the most greedy video pig was reluctant to pay for an endless visual diet of *Goin' Down the Road, Kamouraska* and *Outrageous*. When cable subscribers did sign up for pay-TV, they often didn't stay for the popcorn. In June 1983, for example, 21 per cent of subscribers dropped their First Choice subscriptions from Rogers.

Ted Rogers blamed private broadcasters for the dismal performance of pay-TV. "The private broadcaster lobby delayed the introduction of pay-TV for more than a decade," he said in 1984. "This delay, as much as any other factor, has caused the terrible problems of the past year."

Not only that, lamented Phil Lind, now senior vice-president at Rogers, but the 40 per cent Canadian-content level was "much too high" because there simply wasn't enough Canadian programming available. "Canadian programs get repeated too often, and that's why people express dissatisfaction," he said. "When disconnects are equal to new sales, there's obviously a problem in programming."

There was another reason for the dismal performance of pay TV. Companies like Rogers assumed that subscribers would line up to sign on for the new service, and they didn't bother to promote it. "We didn't sell the need for the service," said the company's director of marketing, Kevin Shea, in 1984. "We just expected people to buy it."

Some people did. By mid-1985, more than 20 per cent of cable subscribers had forked over an extra $10 a month for pay-TV service. That still left another 10 per cent to go before subscriptions equalled the government's rosy predictions of 1982. And VCRs and satellite dishes threatened to make pay-TV a dead dog long before it started to do tricks. But then 20 per cent was better than nothing.

Recessions and setbacks aside, Rogers continued to look for opportunities. In 1983, George Fierheller, now a member of Rogers's team, called from Vancouver to tell Rogers about a new technology that he'd heard about from an acquaintance in the paging industry. Rogers had never heard of cellular telephones, and he wasn't sure how they worked. "But I don't have to understand how it works," he said, "just that it gives a better lifestyle."

So what if Rogers Cablesystems was wallowing in a swamp of debt? This was another technology with great potential, and all Rogers needed was a few million dollars. Rogers's directors saw him coming, and they recognized that familiar glint in his eye. Rogers asked for his directors' approval to apply for a cellular licence and invest a few million in a new cellular service. The board said no. As the controlling shareholder, he could have overruled their decision. Instead, he took a more prudent course. With $2 million of his own money, he formed Cantel as a partnership with Marc Belzberg of First City Financial Corp. in Vancouver and Philippe de Gaspé Beaubien of Telemedia Inc. in Montreal.

With entrepreneurs from three Canadian cities behind the application, the government awarded a national licence to Rogers to compete against the cellular divisions of the regional phone companies. In the process, Rogers beat out eight other applicants, including CNCP Telecommunications, Selkirk Communications Ltd., Abitibi-Price Inc. and Motorola Canada Ltd. "We had no right to think we'd get the licence," Fierheller said. "We weren't a phone company, and there were other applicants with more experience than Cantel."

To the phone companies, Rogers was still a flea bite on the backside of the telecommunications giant. But Rogers and his executives knew that their entry into the cellular industry marked the beginning of their head-to-head competition with Bell and the other Canadian phone companies. Over the next two years, Rogers added another $5 million to get Cantel up and running. Soon, Canadians from coast to coast were chewing the fat from their cars, from airports, even from the sidewalk in front of the shoe store. Everyone, it seemed, had a cell phone. By 1990, Cantel had 200,000 subscribers and a 1 per cent penetration rate. (McCaw Cellular, the leading U.S. cellular company at the time, had only a 0.66 per cent rate.) By 1995, it had almost 800,000 subscribers, with a penetration rate of more than 3 per cent.

By the end of the 1980s, Rogers had acquired control of Cantel from Telemedia and the Belzbergs. "We spent more time arguing about when we would meet than what we would do," he said. "One would be on the phone, one would be late, and the other wasn't there."

Fierheller was more diplomatic. "It's difficult running a company with three entrepreneurs," he said.

In bringing Cantel into the Rogers fold and rolling his five privately held radio stations into RCI, Rogers was building the financial clout of his company and laying the groundwork for a communications empire. He had purchased Multilingual Television (Toronto) Ltd. and its Toronto television station, CFMT, in 1986 from Dan Ianuzzi, hoping to obtain approval for a nationwide ethnic network. But when Rogers applied to the CRTC for approval, the regulators rejected his plan. "Basically, the CRTC didn't license us because the money wasn't there," said Jim Sward, then president of Rogers Broadcasting.

The next year, Rogers launched a bid for Selkirk Communications Ltd., which owned cable networks, fourteen radio stations and three TV stations. But this bid, too, was rejected after Selkirk's dominant shareholder, Southam Inc., refused to sell. Rogers took the rejection with good humour. "I'm not going to do anything to get into an argument with them," he said. Anyway, he had other business to attend to.

In 1989, he acquired the outstanding shares of Western Cablevision Ltd. in Vancouver, of which he already owned 45 per cent. He increased his stake in Astral Bellevue Pathé Inc. in Montreal, which controlled French- and English-language pay-TV channels, including First Choice. He owned a 9.3 per cent share of Moffatt Communications Ltd. in Winnipeg, which he sold later that year. And he had invested heavily in Western International Communications Ltd., a $51-million company that controlled Canadian Satellite Communications Inc. (Cancom).

Under regulations passed during the reign of Pierre Trudeau, before a company could hold a broadcasting licence in Canada, Canadian citizens had to control at least 80 per cent of its voting shares and 80 per cent of its paid-up capital. With this in mind, Rogers had started to reduce the level of foreign ownership in his company. Among other things, he de-listed his company's shares in the United States and bought back shares from foreign owners. He also sold Rogers U.S. cable holdings for $1.63 billion, using $1 billion to pay down his company's debt and the rest to strengthen his Canadian operations.

In the process of all this activity, he alerted the investment community to the possibility that he might have bigger things in mind than delivering televised hockey games and providing telephone service to cars. "I think he's structuring the company for something grander," a prophetic but unidentified analyst told the *Financial Post*. "It appears it's going to be done through Rogers Communications, not RTL [Rogers's holding company]."

Rogers admitted in 1989 to his grand vision in an interview in *Maclean's* with Peter Newman. "What I really want to do is establish a Second Force in Canadian communications," he said. "Between cable and Cantel, we have more than 2 million subscribers, and it would be logical to combine CNCP with the cable operations, so that they lay the fibre optics between the cities, for instance, and we distribute the service to homes and businesses."

"That may sound too visionary to be taken seriously," Newman remarked. But if anybody could make it happen, Ted Rogers could.

At its annual meeting in April 1989, Rogers said his company would spend $1 billion over the next three years to improve its cable and cellular networks and create a 7,800-kilometre cellular network stretching from coast to coast. Analysts took this as a sign that Rogers was preparing to challenge the long-distance monopoly of the telephone companies. "He's setting up the infrastructure that he'll need to provide long-distance telephone service," said Laurel Slocum, an analyst with Alfred Bunting & Co.

Three weeks later, Rogers announced that he would purchase a 40 per cent stake in CNCP Telecommunications. Promising to end "Soviet-style communications monopolism" in Canada, Rogers said he would apply later in the year for permission to become a long-distance provider, competing against a consortium of telephone companies in the $5 billion market. Once again, Rogers anticipated enormous costs in setting up the network, and he predicted that the long-distance company wouldn't make a profit until 1999. But if the prospect unsettled Rogers, it didn't faze his partner in the new company, Canadian Pacific Ltd.

With its previous partner, Canadian National Railways, CP had applied unsuccessfully in 1983 to become a long-distance competitor. Now six years later, with a new, more aggressive and flexible partner, CP expected its second application to succeed, and so did others. "With Rogers's marketing and political skills and the credibility of the organizations behind him, I think he has a fair chance of winning," said analyst Peter Legault. "Twenty years from now, there will be two giants, Bell Canada and Rogers Communications, providing cable and phone service across the country."

Rogers and CP, which had finally found a name for their company, didn't file an application for Unitel Communications Inc. to become a long-distance carrier until the following May. By then, Rogers had received wide and favorable coverage in the North American business press and had put forward his arguments for increased competition, including lower rates and better service, not only in the media but to politicians and bureaucrats through lobbyists in Ottawa. Meanwhile, Unitel president George Harvey toured the country, talking to community groups about the advantages of long-distance competition. And in the Canadian press, the value of the long-distance market grew mysteriously from $5 billion, as it was originally described in the *Financial Times of Canada*, to $7 billion in the *Financial Post* a few months later.

Regardless of the exact figure, the market represented a lot of dimes and quarters. Rogers knew he could deliver better service at lower cost, and he was prepared to fight for the chance. At CRTC hearings in 1991 to examine Unitel's application, as well as another bid from a joint venture involving Lightel Inc. of Toronto and B.C. Rail Telecommunications, a unit of B.C. Railway, lawyers for the phone companies suggested that Rogers could not afford the $13 billion it would take to build a competitive long-distance system. "I'm a builder of companies," Rogers responded. "Just because I'm not as rich as your client, I don't have to apologize."

Another fifteen months was needed for the CRTC to make up its mind, and, when the big day finally came, on June 12, 1992, it took CRTC chairman Keith Spicer more than an hour to ramble through his lofty pronouncement about the information age to the huddled masses of the media who stood before him in utter boredom. But it was worth the wait. When Spicer finally got to the point, he announced that the CRTC would open the long-distance market in Canada to competition. By now, the business press had raised the value of the market to $7.5 billion. With the B.C. Rail joint venture poised to dive into the fray, Unitel pledged to hire five hundred people and spend $100 million on new equipment over the next year. As it turned out, Unitel opened for business just four months later, when Olympic medallist Silken Laumann placed the first public long-distance phone call over Unitel's network to a friend in Victoria before a cheering throng of squinty-eyed TV cameramen and bedraggled scribbling reporters.

Within a year, as the phone companies appealed the CRTC's decision to the federal court and spread the word to Canadian consumers that Unitel was operating on shaky ground, long-distance rates began to tumble, just as Rogers had predicted. Unitel, however, continued to lose money hand over fist, even after selling a 20 per cent stake to AT&T in 1994.

As Unitel struggled, Rogers was attending to other strands in the fabric of his communications tapestry. In 1993, he approached Maclean Hunter to buy its paging operations through Rogers Cantel. Valued at $30 million to $60 million, paging formed only a small part of the Maclean Hunter empire, but it would fit conveniently with Rogers's paging operations.

Then, on February 2, 1994, Rogers made his most audacious move yet. Offering $17 a share plus a payment based on the sale of the target company's U.S. cable operations, Rogers launched an unsolicited bid for Maclean Hunter Ltd., a 107-year-old media organization with a history of conservative, risk-averse but profitable growth. Like Rogers's company, Maclean Hunter had started inconspicuously, with a magazine called *Canadian Grocer*. But while Rogers had taken only thirty-four years to build his empire, Maclean Hunter took more than a century. For decades, Maclean Hunter had been run by accountants who kept the books balanced and the corporate boat on a steady if unspectacular course. In magazine publishing, the company was renowned as a follower, not a leader, waiting until the 1980s to promote a woman beyond the mid-management ranks and waiting even longer to acquire its first editorial computer terminals. However, Maclean Hunter had succeeded in building a solid cable TV operation, and it now ranked fourth in the country.

Their corporate colours reflected the two very different organizations. Maclean Hunter puttered around in vans painted an innocuous shade of blue that resembled a polyester golf shirt from Kresge's, lettered in unremarkable drab tones of grey and gold. On a downtown street, a Maclean Hunter van could quickly become lost amid the thousands of plumbers' trucks and tacky courier vehicles all painted the same general colour, which could be mixed in a swimming pool for 45¢ and touched up with spray paint from Canadian Tire. The blue and grey and gold were safe, unthreatening and humourless, the sheen subdued and the twinkle removed, like someone's Presbyterian grandmother on Valium. Meanwhile, Rogers raced from call to call in vans painted fire-engine red. The motorized equivalent of the Solid Gold Dancers, they broadcast sex appeal and spirit and polished aggression. Rogers red was unmistakable: If it wasn't a Coca-Cola truck, it was Rogers.

Ted Rogers, who had quietly accumulated 8 per cent of Maclean Hunter's shares at about $12 or $13 apiece, called his offer an attempt to form a strategic alliance. Maclean Hunter managers called it "a takeover attempt." No matter what you called it, the acquisition would be the largest merger in the history of Canada's communications industry and would give Rogers about 35 per cent of Canada's cable industry.

As usual, Rogers had to justify the acquisition to regulators, rivals and the Canadian public, and he began before the offer was even a day old. Only a company of substance can compete in the interna-

tional arena against satellite and cable conglomerates like Time
Warner and Bertelsmann, he argued. Total revenues of Maclean
Hunter and Rogers combined would amount to $3.1 billion, com-
pared to Time Warner's $18.4 billion, Bertelsmann AG's $13.9 billion
and News Corp.'s $9.5 billion. "These new international conglomer-
ates are already large enough to dominate the Canadian marketplace
without effort," Rogers continued. "Small Canadian companies will
not be able to compete even for local audiences, let alone the attention of the rest of the world. These global conglomerates will not look
to Canada and Canadians for technological innovation, either."

Besides, Rogers added, "with the two companies put together, we
can do more things in R&D and develop more fibre technologies than
we can as separate companies."

Media conglomerates weren't Rogers's only concern. He was worried about the phone companies, as well. "If the cable industry is to
have a chance of being competitive against the voracious advances of
the telephone companies, then cable companies must be in a position
to consolidate," added Phil Lind.

In the opinion of most analysts, the bid was fair. After all, Maclean
Hunter shares had traded in the range of $12 to $13 a share for the last
five years. "It's a reasonable bid," observed Frank Mersch of Altamira
Investment Management.

With the raging barbarians just outside the ramparts, Maclean
Hunter's management called a board meeting for the following day,
February 3, and urged its financial advisers, RBC Dominion Securities
and Wood Gundy, to assess the bid. Meanwhile, the company began
shaking the corporate bushes for competing bidders.

Politicians and regulators also sensed a blessed opportunity for
self-promotion and jumped into the fray. "We want to make sure the
fundamental elements of competition are preserved," blathered John

Manley, the industry minister, in Ottawa. The federal Bureau of Competition Policy had the power to unwind the deal if it found that the merger would reduce competition, he claimed.

By February 16, Maclean Hunter and its advisers had reached a conclusion. In a news release, they called Rogers's bid "inadequate" and said they would use the company's poison pill, which would make Maclean Hunter far less attractive as a takeover target by diluting the value of its shares. The company also said it would try to sell its Canadian cable operations or form a strategic alliance with a third company. In response, Rogers said that even his wife couldn't make him raise his bid above $17 a share. Then he packed his socks and Bermuda shorts in his briefcase and left for a cruise on his yacht in the Caribbean.

To accommodate rival bidders, Maclean Hunter set up data rooms in Toronto where potential suitors could review the company's books and question managers about specific details of the company. To enter a data room, a visitor had to sign a confidentiality agreement, which included a stipulation that the potential bidder would take no action for 120 days. As it solicited new dance partners, however, Maclean Hunter excluded Rogers from the party rooms. The company's CEO, Ron Osborne, compared Rogers to "a Detroit knee-capper." Miffed, Rogers took Maclean Hunter to court. In a rare Sunday hearing of the Ontario Court's General Division on February 27, Rogers lawyer Robert Armstrong likened the situation to the sale of a house. "[Maclean Hunter has] hired a real estate agent," he said. "They've got a For Sale sign on the house. They're having an open house. How can they ignore Rogers?"

With all the wit he could muster in a room full of lawyers on a Toronto Sunday morning in late winter, Mr. Justice James Farley responded, "I guess they want to have a warm feeling that you'll look after the house and that you won't paint over the panelling in the den." The next day, Farley rejected Rogers's application for access to Maclean Hunter's data rooms. He also ordered Rogers to pay $60,000 in court costs that Maclean Hunter had incurred defending the action.

Obligated to its shareholders to sell the company's assets for the highest price, Maclean Hunter continued to look for competitive bidders. But everywhere it turned, Rogers seemed to have arrived sooner. Shaw Communications Inc. in Edmonton, the country's third-largest cable operator, might have jumped in to bid on Maclean Hunter's cable systems, but James Shaw Sr., the president, said on February 18 that "we don't have that kind of borrowing power" to make a bid. Two weeks later, on March 4, Shaw Communications agreed with Rogers to swap cable systems if Rogers acquired Maclean Hunter. Rogers had also contacted Telemedia Inc. and Quebecor Inc., both in Montreal, and both potential bidders for some of Maclean Hunter's assets.

With nothing else to do but wait, journalists tried to create some action of their own. Rumours began to circulate that BCE, which was already preparing to buy a 30 per cent stake in a California cable company called Jones Intercable Inc., might urge Jones to buy Maclean Hunter's U.S. cable assets. BCE itself was identified as a potential white knight, along with Power Corp.

In Ottawa, the federal government made life more difficult for Maclean Hunter when it closed a tax loophole that would have allowed the company to sell its U.S. cable assets and defer capital gains tax while passing hundreds of millions in tax savings along to shareholders. And in the United States, where cable rates are at least twice as high as they are in Canada, regulators reduced by 7 per cent the rates for basic cable service. According to news reports, the reduc-

47

tion could have affected the value of Maclean Hunter's U.S. cable operations.

Still confident that his bid would succeed, Rogers continued to pull the levers of his public relations machine to win support of Canadians and their self-appointed watchdogs. In February, he hired Pierre Juneau, former president of the CBC, former chairman of the CRTC and former federal deputy minister of communications, to act as trustee on his behalf to guide Maclean Hunter while the regulators scratched their heads and wondered if they would approve the deal, a process that could take a year or more.

On Sunday night, March 6, Ron Osborne took a long drive over a few short blocks through midtown Toronto to visit privately with Ted Rogers at his house in Forest Hill. Maclean Hunter was stymied, he would later admit. The company simply couldn't find a better deal.

Two days later, as analysts and investors awaited the opening salvo of Rogers's annual meeting at the Art Gallery of Ontario, they wondered among themselves if Rogers would extend its offer beyond the March 15 expiry date. U.S. regulators also had to give their blessing to the deal, since it involved U.S. cable operations, and they had given Maclean Hunter and Rogers a few days to submit their arguments. Others wondered if Rogers was waiting for Maclean Hunter's board to make a public statement before extending the expiry date. But Phil Lind said that wasn't the case. "We're just holding off until tomorrow because we want to be sure that when we say the thing, we say it right." Analysts nodded and wondered what in the world he was talking about.

That afternoon, however, as the annual meeting proceeded, a lawyer from Tory Tory DesLauriers & Binnington rushed into the gallery's central court and passed a piece of paper to Rogers chairman Garfield Emerson. As if it were a hot potato, Emerson passed it to Ted

Rogers, who then showed it to Phil Lind. On the podium, meanwhile, one of Rogers's financial executives continued to deliver a long-winded presentation on the company's financial performance. A flurry of scribbling ensued involving Rogers and Ron Osborne. Then Rogers stood up to announce that they'd just consummated the deal, right there in public view. The audience of jaded investment professionals and shareholders rose to its feet and applauded. "We were sitting there hoping the reports would go on and on because we didn't know if the lawyers would come up with the final papers on time," said Phil Lind. Rogers had even prepared two speeches for the event, one announcing the deal's completion, the other explaining why the bid's expiry had been extended.

Under the terms of the deal, worth about $3.1 billion, Rogers stuck to his original offer of $17 a share, but threw in 50¢ a share in the form of a dividend paid by Maclean Hunter. Rogers paid $3 billion for Maclean Hunter and assumed about $490 million of outstanding Maclean Hunter debt. Until Rogers could sell Maclean Hunter's U.S. cable operations, he was carrying a $2-billion debt, including a $1.5-billion loan.

Upon approval, however, the deal would give Rogers control of most of Toronto's cable service as well as such Maclean Hunter assets as its newspapers and periodicals, including *Maclean's* and the *Financial Post*, radio and television stations and a state-of-the-art printing plant in Aurora, north of Toronto. "I'm just sorry Maclean Hunter doesn't have a movie studio," he said. "But I guess we can't have everything."

Predictably, the usual gaggle of critics gathered round the wailing wall to gnash their teeth and knit their brows over the impact of the deal on the future of Canada. "It's a bad thing for the world of ideas," pouted Ian Morrison, president of the Friends of Canadian

Broadcasting. "What you're getting is something that approaches mind control."

"It's not visionary and forward thinking; it's greedy," lamented Professor Hudson Janisch.

Most investors, on the other hand, applauded the deal and gave Rogers full marks for his achievement. "Twenty years ago, the idea that Rogers would take over Maclean Hunter would be given the same probability as Joe's Hot Dog Stand taking over McDonald's," said investment manager Ira Gluskin.

As for Ron Osborne, he lamented the passing of Maclean Hunter into the hands of an aggressor. "I'm as sorry as anybody that this transaction results in the demise of a fine Canadian establishment," he said. "And I take no pride in that demise. We fought like hell not to have that demise happen back in February and March, and I make no bones about it."

Once the deal was completed, Rogers still had more obstacles to overcome. Regulators and the Canadian public alike had to be convinced that the deal would contribute to Canada's future. "Strategic mergers occur when old conditions governing a marketplace are swept away by a new force," Rogers announced in a glossy twelve-page colour brochure to explain the merger. Over the next six months, with Rogers in action, the CRTC received 740 letters about the Maclean Hunter acquisition, of which only twenty opposed it.

On September 21, Rogers and his team of Garfield Emerson, Phil Lind and Graham Savage, along with Maclean Hunter CEO Ron Osborne, appeared before the CRTC. Chairman Keith Spicer and Paul Temple, CRTC vice-president of regulatory and public affairs, both grilled Rogers for several hours about rates, service and other issues. "You've got a God-sent opportunity to lower rates," Spicer said at one point. "I think you should take advantage of that opportunity to tell

Canadians you'll lower rates. Would you like to commit today that you'll give subscribers back half the savings?" he asked.

"No," Rogers replied.

To substantiate his claim that the deal would benefit all Canadians, Rogers committed to spending $102 million over the next five years to provide households with better access to the "electronic networks of the future." According to his critics, at least half of this spending would have occurred anyway as Rogers upgraded equipment and installed new technology. But Rogers also pledged to spend $4 million to support "the great tradition of Canadian documentary filmmaking" and $3.9 million for community programming, which, after all, gave Canadians such luminaries as Micki Moore and Ed the Sock.

As part of the acquisition, Rogers gained a one-seventh interest in the CTV network through his ownership of CFCN Communications Ltd. in Calgary. After asking Ron Osborne to represent Rogers on CTV's board, Rogers criticized the network, saying it would "go down the tubes" unless its owners restructured their holdings. He said they needed to eliminate conflicts among CTV members who also own independent broadcasting outlets that compete against each other. "He has never been to a CTV board meeting and he has never been involved with CTV and I think he should hold his opinions until he does so," replied Douglas Holtby, chief executive of Western International Communications Ltd. in Vancouver.

To reduce his $2-billion debt, Rogers had started to unload assets, including Maclean Hunter Printing, sold to Quebecor, and the company's publishing operations in the U.K. In June, he sold the company's U.S. cable interests to Philadelphia's Comcast Corp., which employed Rogers's son, Edward, at the time, for about $1.76 billion. Combined with sales and swaps of other assets, Rogers raised enough money to pay off its $2-billion debt before the end of the year.

In December, the CRTC approved Rogers's acquisition, with only one notable qualification: It demanded that Rogers sell CFCN in Calgary. Rogers said he was horrified at the prospect, but it also removed him from a potential wrangle with CTV's board, some of whose members he had already alienated before setting foot in their boardroom, and relieved him of the delicate task of finding a job for Ron Osborne. With little role to play in the company's affairs now that he could no longer represent Rogers at CTV, Osborne soon accepted a position with BCE.

In buying Maclean Hunter, Rogers had staked his family's entire holdings, worth about $1.2 billion, to complete the deal. He now headed an operation that employed 13,000 people. And there was more to come. "This isn't the end of mergers of cable companies," he predicted in 1995. "I think you'll see more of it."

CHAPTER 3

The Technologies

I have seen the light

Television enables you to be entertained in your home by people you wouldn't have in your home.

DAVID FROST

Like Ted Turner, Bill Gates, Richard Branson, Ben & Jerry, and other successful entrepreneurs, Ted Rogers does not invent new technologies, he exploits them. In this capacity, he excels. From cable to cellular, Rogers may not have been among the first to invest in new technologies, but he has been the first to exploit them to their maximum potential. And even as he dreams of the multitudes using a Rogers digital dingus to record, manipulate and transmit every thought – exchanging faxes, voice messages, old movies, even digital photographs from street corners, automobiles and living rooms – inventors are finding new ways to help him fulfill his dream.

Rogers and the nerds who provide him with his raw technology live in different worlds. Inventors beaver away for days on end in cluttered, windowless laboratories, staining their fingers with chemicals and poking at wires with needle-nosed pliers while monitors flicker in the background. At the end of a long session in the lab, the last thing an inventor wants to do is join a clutch of anal-retentive bankers polishing their neatly clipped fingernails on their lapels in a sixty-fourth-floor boardroom above Toronto's financial district while they haggle over the details of a $500-million bond issue. He'll leave that excitement to guys like Rogers.

As an entrepreneur, Rogers may take an interest in a technology, but he's far more interested in turning it into a money-making proposition. The most influential entrepreneur in Canada's communications industry, Rogers has seldom used a computer, nor does he spend more than a few minutes a day watching TV. But he knows what people will buy. They bought FM radio in the 1960s, cable TV in the 1970s, cellular phones in the 1980s, and fibre-optic technologies in the 1990s. And as phone companies, cable firms, equipment manufacturers and their purse-lipped regulators patch together a digital universe like farmers' wives at a quilting bee, we'll likely end up buying information of all kinds, as well as access to it, from a Rogers-owned company, well into the twenty-first century.

Rogers has had nothing to do with inventing radios, televisions, cellular phones or the networks that join them together, but he was far ahead of the pack in selling them. "I believe in getting in on new technology in the formative stages to ride the wave of growth," he said. "Even if you make a lot of mistakes, growth covers them up. If you survive, you can do extraordinarily well."

Surviving for more than thirty years, Rogers now controls the leading cable company in Canada, virtually equipped to deliver two-way service to its 2.5 million subscribers in the country's three largest urban markets through one of the most architecturally advanced cable networks in the world. He has the largest single cellular operation in Canada, with 43 per cent of the national market, relying on its own digital microwave and fibre-optic transmission equipment to carry calls across Canada. With the acquisition of Maclean Hunter, he expanded his cable network, gained control of Canada's pre-eminent magazine publisher and 25 per cent of the country's paging operations. His investment in Unitel, the country's second-largest long-distance provider, has not paid off yet; but technology is not to blame for its struggles. Meanwhile, the radio station that he bought in 1960, CHFI, leads the country in audience size and sales, his other radio and television stations perform well, and his home-shopping and multicultural channels have the potential to become a national presence. Whether or not we watch, read or listen to any part of it, Rogers's empire can hardly be ignored.

Engineers have played a significant role in building his companies, and Rogers has always paid close attention to their ideas, even if he doesn't always understand what they're saying. Throughout the 1980s, Rogers would convene a brainstorming session every couple of months with his team of techno-geeks, at which conversations would inevitably turn to technology and the future. As the techies poked at their chicken and knocked back wine by the tankerload, Rogers would listen to their excited comments about multiplexing, compression, bandwidth line extenders and switching systems. "I open my meetings with the engineers," he said, "by asking 'What's new in these trends?' In many cases, I don't know what they're talking about. But if everybody nods, I nod."

That doesn't mean he's a technological idiot. He knows the basic principles, and he understands the application of the sync generators, optical time domain reflectometers, multiplexers, demultiplexers, digital modulation analyzers, oscilloscopes and transmitters that make his company tick. During negotiations to acquire a cable-TV outlet, he once redrew a diagram that appeared within the bowels of a hundred-page legal document because it had misrepresented a transmitter pattern. But technology itself doesn't motivate Rogers as much as its potential impact on the way we live. "I don't innovate technologies as much as new lifestyles," he said. "The technology, whatever it is, is just the instrument."

The technology to which Rogers and every other broadcasting and communications mogul in the world today owe their fortunes has been the work of a pantheon of inventors and tinkerers. Most of them died before Rogers was even born, but if it's any consolation, their names endure on radios, telephones or measurements of sound. They include Samuel Morse, who sent the first long-distance telegraph message in the United States in 1844; Alexander Graham Bell, who invented the telephone in 1876, founded *Science* magazine, sent the first wireless telephone message, introduced Helen Keller to Anne Sullivan and patented the photophone, which used sunlight and a special light-sensitive device in the receiver to relay and subsequently reproduce a human voice; and Guglielmo Marconi, who put St. John's, Newfoundland, on the map in 1901, when he sent the first transatlantic wireless telegraph to Signal Hill from Cornwall, England.

The most influential historical figure in this team of electrically charged inventors was Heinrich Rudolf Hertz, the prodigious wizard who first demonstrated the existence of electromagnetic waves in 1888. People had talked about the properties of sound and light for years. But no one had actually proven their existence, mainly because no one could figure out how to measure, control and monitor waves that they couldn't see and that moved faster than you could blink. For most of the nineteenth century, leading scientists stroked their beards, puffed up their chests and postulated that waves were elastic in nature and that they travelled through a kind of cosmic tar, invisible to the eye and unsuitable for paving a driveway, but real, just the same. (With a similar sense of irrefutable logic, they also postulated that a person's character traits could be measured by mapping the bumps on his head.)

In the 1870s, James Clerk Maxwell, a Scottish physicist immortalized by the Beatles in a song about his silver hammer, developed a set of equations that unified electricity, magnetism and light. These equations sent the puffy-chested pooh-bahs of science into apoplectic fits. They gave Maxwell credit for designing a nice set of equations. But the proof was in the pudding, or the cosmic tar, and for ten years no one could come up with a suitable experiment to validate Maxwell's equations. Then along came Hertz.

As a struggling unpaid lecturer in Kiel, and later as a professor of physics in Karlsruhe, Hertz turned his mind to Maxwell's equations. But even with the encouragement of his mentor, Hermann von Helmholtz, another German with a name like a *Batman* character, Hertz could not think of a procedure that would effectively demonstrate the relationship between electromagnetic actions and the polarization of a dielectric, which is what Maxwell's equations are all about, if you really want to know. For one thing, Hertz had to measure the speed of electric waves with existing equipment, a task he compared, chuckling through his schnitzel, to "marking consecutive lengths of an almost microscopic object with a piece of chalk." To improve on the electronic chalk that was then available, Hertz devised his own instrument, which physicists still refer to as the Hertz oscillator. He also devised a formula for detaching electric and magnetic fields from wires, which enabled them to go free as Maxwell's waves. The Hertz vector is still applied in the operation of transmitting antennae and other communications technology.

To prove conclusively the existence of electromagnetic waves, Hertz dragged a flat sheet of zinc the size of an eiderdown blanket into his lecture hall and mounted it on a wall. Then he placed his oscillator against the opposite wall and turned on his machine. Like a mechanical pitcher in batting practice, it lobbed electromagnetic waves at the zinc sheet. The waves bounced off the sheet, and the room became a disco inferno of undulating pulses. Using a crude receiver consisting of a pair of metal rods placed end to end with a small gap between them for a spark, Hertz moved about the room, locating points where waves, bouncing one way off the zinc sheet, reinforced or opposed the waves being hurled the other way by the oscillator. In the process, he validated Maxwell's theory. He also verified Maxwell's prediction about the speed of the waves – the speed of light. Though Hertz didn't know it at the time, his experiment would eventually make it possible to play Huey Lewis and the News on your car radio, watch Crazy Guggenheim make faces at Jackie Gleason on TV, or call the White House on your cell phone.

For his efforts, the Berlin Academy of Sciences gave Hertz a mittful of deutsche marks, which today would hardly cover his monthly cable bill. But his name lives on to this day, immortalized as a measure of frequency – the number of waves that pass a specific point in one second. One wave, or cycle, per second equals one Hertz. Unfortunately, Hertz himself didn't live on much longer. He died in January 1894, a few weeks before his thirty-seventh birthday.

Because of Hertz, it became possible at last to regard information as a commodity. Through Hertz's efforts, you could now create packages of information, neatly wrapped bon-bons of aural and visual delight, convert them, load them onto electromagnetic transport vehicles, transmit them to their destination and receive them at the other end. Like confectioners creating and shipping Godiva chocolates or iron manufacturers delivering slabs of metal, operators of radio, tele-phone and television companies would eventually acquire the equip-ment to package their distinctive electromagnetic products and hawk them to consumers, for a price.

Until Hertz came along, air was just dead space. After Hertz, peo-ple realized that the air was full of waves, bouncing and undulating in patterns like a mob of acrobats and trapeze artists dressed up as mid-level managers in herringbone suits and forming a spectrum from the slowest moving to the speediest, including microwaves, light waves, X-rays and cosmic rays. The waves themselves could perform all sorts of tricks. If you could figure out how to capture and control them, you could create the world's greatest circus of sound and light, or cook your chicken dinner in two minutes. As Michael Mirabito and Barbara Morgenstern write in *The New Communications Tech-nologies*, "A signal can assume an infinite number of variations in terms of its amplitude and frequency, within the operational bounds of the communications system."

Once you could identify and manipulate their properties, each wave could conceivably carry bags of information to its destination. There, your suitably equipped radio or television or telephone receiv-er would open the bag, and you could express appropriate shock, horror, delight or boredom over the contents. For little bags, there are little waves; for big bags, big waves. You could even trick the little waves into carrying big bagfuls of information by compressing the equivalent of a steamer trunk into something the size of your wallet. To make them move more quickly or carry more baggage, you could pump up the waves with electronic steroids. You could choreograph the waves to dance together, or you could let each wave do its own thing in anarchistic abandon.

For people like Paul Reuter, founder of Reuters news agency, and

Walt Disney, founder of Tomorrow Land, the ability to package sound and images, duplicate the packages and send them through the air was a godsend, and they amassed huge fortunes in the process. For lesser-known lights like Edwin Armstrong, who devised a method of modulating the frequency of sound waves in the 1920s, or R. Gordon Gould of Columbia University, who patented a laser in 1957 only to have his patent classified and his notebooks confiscated by the U.S. Defense Department, the joy lay in developing the equipment that could manipulate the waves and make them more productive by focusing their power on a specific task.

In the early days of Canada's communications industry, a wave was a wave. You created a wave at one end of the process and converted it into a magnetic or electronic pulse, analogous to the original wave. Then you transmitted the pulse over a wire or through the air and duplicated the wave at the other end. Companies like Bell Telephone and Rogers Batteryless were built on analog equipment that did just this.

In the process of recording or transmitting a signal, the electrical properties of an analog wave, as Heinrich Hertz confirmed, can be manipulated. By passing the signal through a transducer, such as a microphone, the sound can be carried on a higher-frequency carrier signal, transmitted to a receiver, then converted back to its original wave pattern. Or a vinyl record can be impregnated with a charge, which it delivers to the tone arm of a record player. Translated from a pulse into a continuous electrical wave, the signal then passes through an amplifier, activates your speakers and keeps the neighbours awake. Regardless of the equipment involved, the analog wave that goes in is identical to the one that comes out, unless it gets distorted along the way, which happens all too often, especially when you're driving through the Berkshires at two o'clock in the morning on your way to

Boston to catch a Red Sox game, listening to WABC from New York on the car radio.

Most of the radio stations that you hear in the mountains at 2 a.m. come from, appropriately enough, AM stations. To provide you in your Toyota with a relatively precise impression of a long-haired hep-cat pounding on a set of drums in a studio, these stations broadcast signals whose strength is modulated according to the pressure exerted on a diaphragm by the sound. Bang the drum loudly, and you get a loud noise in the front seat of your car, carried by a wave as high as the Empire State Building; bang it softly, and a demure but determined little wave in peau-de-soie toe shoes dances over the air with an amplitude slightly higher than a Cartier rolling ring. (The intensity of the sound – the volume, basically – is measured in decibels, each one-tenth of a bel, named after Alexander Graham Bell.)

Although they can travel long distances, bouncing off high buildings and tree-covered mountaintops from Chicago to Huntsville, Ontario, all sorts of things can interfere with an AM signal, from airplane noise and sunspots to a police radio. You can drastically reduce the noise that distorts a signal by modulating its frequency rather than its amplitude, as Ted Rogers realized when he bought his first FM radio station. For reasons best left to professors and radio repairmen to explain, modulating a signal's frequency enhances the fidelity of the sound and enables you to transmit sound in stereo. But FM signals don't travel as well as AM signals. In big cities, where advertisers or the government can reach lots of people who own suitable radios, FM broadcasting makes sense. It made less sense in the 1960s, before cable, satellite and other technologies, to build an FM station in Fort Chimo, Quebec, on the shores of Ungava Bay, because the FM signal would reach only a few walrus hunters gathered around the wireless at the Hudson's Bay post before it floated off into the ozone.

When Rogers bought CHFI in 1960, FM radio had been lurking in the shadows of the broadcasting industry since the 1920s. But it took two important factors to make FM radio a commercial success in Toronto. First, because of its limited scope, an FM radio signal had to reach a concentrated audience. Second, the audience had to have the necessary equipment to receive the signal and turn it into comprehensible sound. In 1960, about 500,000 people lived in Toronto. To receive an FM signal, 95 per cent of them had nothing more receptive than an egg beater, a coat hanger, or a size-7 Hush Puppy. Fortunately, about 25,000 Torontonians, stalwart United Empire Loyalists all, owned FM radio receivers, and Rogers made it easy for others to join the FM club. That was enough to get Rogers started.

After more than thirty-five years in the industry, Rogers no longer has to hand out radios to potential listeners. But he continues to expand and improve his radio operations. In four Canadian cities, he owns both an AM and an FM station – twin sticks, as they're called in the business – which helps the company to consolidate its overheads and weather the perennial ups and downs of radio advertising sales. Meanwhile, he pays attention to their technological well-being, again to ensure that they attract as much advertising revenue as possible. "If you work terribly hard," he said, "you get more capacity and more power and more coverage, and therefore your ratings are higher" – and advertisers clamour to reach your additional listeners.

At their most basic level, radios, telephones and television sets still operate in an analog environment. They still depend on the exact duplication of a wave to convey sounds and images from one point to another. And it requires different equipment to handle different forms of analog information. The signal from a television camera, for example, can't be translated and reproduced by an analog telephone system, nor can you conduct a phone conversation with your brother-in-law about why you can't lend him $500 over your 54-inch Sony TV plugged into the cable outlet, even if you wanted to. Like chickens, omelettes and a dozen eggs, each form of communication may be related to the others. But each requires its own particular equipment built specifically to process, package, distribute and receive it. Just as you wouldn't load a chicken and a dozen eggs in the same carton, stack it in the back of a truck and expect them both to pop out intact at the other end, you couldn't transmit a moving picture of John Travolta and a spreadsheet from the accounting department over the same system – at least, not until recent years. That's because most conventional systems manipulate analog signals to record and receive sounds and images, and the electromagnetic waves created by John Travolta require much different and more extensive treatment than the ones created by an income tax form.

But, to belabour this metaphor until it lies in a puddle half-dead and panting at the reader's feet, if you could reduce the chicken, the omelette and the carton of eggs to their molecular constituents, you could pack all the molecules together and load them on the same slow truck to Regina. They could be reassembled into exact reproductions of chickens, eggs and omelettes at their destination as long as you followed the molecular patterns. That's what happens when information is reduced to its digital form, except that information consists of only two fundamental digital molecules. The informational equivalent of a molecular structure, a digital signal is non-continuous. It doesn't depend for its integrity on the exact transmission and reproduction of a wave. Instead, it depends on the high-speed sampling of a signal – hundreds of thousands of times a second – made possible by microchips and other dazzling wonders of the information age, and the representation of the sampled signal by the digits 1 or 0. To hear Luciano Pavarotti singing his favorite hits from *The Barber of Seville*, you no longer have to send exact reproductions of the peaks and troughs created by Pavarotti's voice in a studio via radio or phonograph record and hope they come out of your speakers without too much distortion. You just have to unload the 1's and 0's in all their splendour, just as they were initially encoded when Pavarotti was bellowing his lungs out, send them in pulses over the airwaves or down a fibre-optic line and your CD player will convert them into music.

Like Morse code, a digital signal consists of a series of simple 1's and 0's – dots and dashes, ons and offs – to represent a sound or image. But instead of telegraph operators poking away at a little button, today's digital systems use microchips to process and transmit 1's and 0's at what seems like the speed of light. A digital system processes and transmits discrete sets of instructions that tell the chips in your telephone or television or CD player to turn themselves on and off, over and over again. By following the instructions faster than you can sing "Gimme a ticket to an aeroplane," your digitally equipped receiver presents information in the form of music, movies or computerized charts and graphs, identical to the original version. The same electronic principles apply to the communication of any digital code, whether it represents a chicken, an egg, huevos rancheros, the Box Tops or Demi Moore. And once the signal is digitized, it can be processed, stored and manipulated by the microchips in a CD player, a digital TV or a computer without losing its integrity.

The digital universe has been evolving for years, and Ted Rogers started to prepare for it long ago. Initially, digital equipment affected operations such as trunk lines and switches, far beyond the consumer's gaze. More recently, engineers have designed digital components for consumer goods such as video recorders and television sets, but it will take a few years before these become generally available at a reasonable price. In the meantime, digital and analog equipment are not usually compatible. So companies like Rogers use specially designed gizmos to carry out analog-to-digital (ATD) and digital-to-analog (DTA) conversion, while viewers remain blissfully unaware that anything is going on at all.

When Rogers first entered the cable industry in 1966, the digital universe was virtually unknown territory. Rogers had been dabbling for a couple of years in television broadcasting with his friends the Bassetts and the Eatons. Television was fun, and it was a growing industry. But TV stations were heavily regulated, and they depended for their profits on the quality of their programming, which could range from a left-handed juggler to *Captain Mike and Buttons* to a Jerry Lewis telethon. It seemed far more profitable to own the distribution system. People might choose one station over another, just as they might choose a Chevrolet over a Ford. But in the 1960s, they had only one decent road to travel, and the road would belong to Rogers. At the time, though, even Rogers had only vague notions that his road might evolve into a highway.

True to his fashion, Rogers entered the cable industry long after the technology had been proven, but long before anyone had figured out how to make much money from it. When he started his cable operation, Community Antenna Television (CATV) systems had been in operation for more than fifteen years. A television dealer in Mahoney City, Pennsylvania, started the first CATV system in 1948; another was set up around the same time in Astoria, Oregon. Both were designed simply to improve the quality of signal reception from local and regional stations. Like fm signals, TV signals could not travel far without breaking up. Carrying far more information than a radio signal, they were also subject to interference. On a clear day, a TV set in Toronto might pick up a jiggly signal from Buffalo. But in a late-night thunderstorm, you might as well play Parcheesi as watch TV. Cable was designed to address this problem.

In a cable system, a programmer sends a signal via satellite, ground transmitter or, more recently, fibre-optic line, to the cable company's receiving equipment. Through an installation called a headend, the cable company then amplifies and transmits the signal through its cable to the consumer. Along the way, the signal is poked, prodded, processed, compressed, adjusted, expanded and monitored by specially designed and often outrageously expensive equipment.

In 1966, cable's Cinderella was still toiling in the scullery, while its ugly broadcasting stepsisters received all the glory and attention. In South Africa, the government thought so highly of television that it prohibited anyone from owning a TV in 1959, a law that remained in force until 1975, long after *My Mother, The Car* was off the air and could no longer corrupt the minds of South Africans. No one in South Africa or anywhere else gave much thought to cable, though. In Canada's newly minted Broadcasting Act of 1968, cable wasn't even considered to be a technology of influence. Restricted to carrying local programming, it was just a glorified town hall, where guys in ill-fitting suits could blather away to their heart's content about residential zoning laws to an audience of three retired schoolteachers and a sheep dog. Rogers received his first cable licence from the federal Department of Transport. Among its activities, the DOT, now Transport Canada, administers the licensing of airport maintenance crews, who have to use their two-way radios to get permission from an air-traffic controller to cross a runway. At the time it also administered licences for cable TV. As long as you had $25 and could recite your Alpha, Bravo, Charlies, you were OK with the DOT.

At first, Rogers was restricted by law from transmitting anything but local fare like *Razzle Dazzle* from Toronto stations. Canadian cable's first big break came in 1969, when the CRTC allowed operators to receive and transmit signals via microwave technology from more distant stations in the United States. Based on the number of cable subscribers, companies like Rogers paid broadcasters a fee to carry their signals. Instead of watching Mayor Pitkethelly exercising his adenoids, cable subscribers could now watch more sophisticated programming such as the *Beverly Hillbillies* and *Shindig*. For such bracing fare, subscribers seemed more than willing to pay $4 a month or more for their cable service, especially when Rogers and other cable operators added new channels to their menu. In fact, so many people wanted better TV reception that by 1991, when the Broadcasting Act was revised, more than 70 per cent of Canadian households had subscribed to cable.

Subscriber revenue allowed cable companies to pay programmers for their shows and still make a profit. Today, cable companies pay wholesale programmer rates starting at zero and rising to 60¢ or more per viewer per month. TV5, for example, a French-only service produced by Canadian and European broadcasters, receives 5¢ per subscriber per month; CBC Newsworld receives 59¢.

Some operators saw a chance to stash the rest of their profits into the pockets of their Tip Top suits. They didn't last long, however, because other operators, like Rogers, soon hung them out to dry, suits and all, quickly taking advantage of new technologies to improve their services, making even more money in the process and taking over their competitors. To pay for continual improvements to their systems, cable operators not only take their cut from the basic cable fee, which is regulated, but also charge subscribers a monthly fee of $3 to $10 for such things as pay-TV, descrambling equipment, optional cable channels and additional outlets. (Unlike phone companies, which add extra lines at no charge except installation, cable companies argue that their outlets deliver more value. "We provide more value in an extra outlet," explained Rogers Cablesystems president Colin Watson at a CRTC hearing in 1994. "With two telephone outlets, you can still have only one conversation at a time. With two cable outlets, you can watch a different program on each outlet.")

In demanding these extra fees, cable operators have been portrayed as rapacious bandits, often by the telephone companies that would like a piece of their action. But the phone companies would find ways to line their own wallets, if regulators would let them. In 1983, for example, a regional telephone company in the United States tried to charge modem users a fee of $49 a month for linking a computer to the telephone line. (A modem converts a computer's digital codes into an analog pattern for transmission along a phone wire.) Phone companies also wanted to charge an access fee of several dollars an hour to users of Dow Jones, the Source, and other information services, on top of regular phone rates. The suggestion was dropped after U.S. customers threatened to make telephone regulators sit in front of a Reuters screen staring at stock quotations for three months or until they went blind, whichever came first.

Meanwhile, Rogers was exploring other ways of satisfying Canadians' insatiable need to hurl signals at each other through the ozone, and many of them brought him closer to competing directly with the telephone companies. His cable service had already provided many homeowners with the first glimpses of an alternative to their telephone line as a conduit for information. Now he saw another alternative that involved no lines at all. In 1982, to the dismay of his own board, Rogers co-founded Cantel to deliver cellular telephone services. Cellular technology, which had been developed around the turn of the century, enables subscribers to dial into the public switched telephone network while scooting around a city where coverage is provided. Initially, one enormous mobile phone tower radiated its signals over a large area such as Toronto. But only twenty-five channels were available to the public, so only twenty-five people could use their cell phones at the same time. This was of limited value in a city of 1 million people, all with something important to say. Cantel solved this problem by erecting a large number of low, weak towers to replace the single jumbo tower. This allowed many people using analog and, later, digital instruments, to travel through cells of the city, each served by a designated transmitter that sends and receives phone messages over the air. The static and interference that occurred when a customer passed from one cell to another was eliminated by technology that could hand off a call from the area served by one tower to the next. Call following, another major step forward in cellular technology, made it possible to call a Cantel subscriber almost anywhere in the country that's serviced by a cellular system. In fact, thanks to Rogers, Canada was ahead of the United States in cellular service, where call following was not available until 1991. Meanwhile, with digital technology, there's even more available air space in each cell for a call, making it possible for that guy at the next table in the restaurant to blab into his cell phone while he's scarfing his vermicelli and squid.

Within a decade, Rogers was envisioning a phone network for voice and data that extended far beyond the confines of a single city. By 1989, he was discussing the idea of a "Second Force, which will be licensed for long distance and data transmission across the country" with his former schoolmate, *Maclean's* columnist Peter Newman. "All of us one day will have portable phones, so there won't be any need for wires all over the place," he added.

Full long-distance service, however, would require more than Rogers himself could offer at the time. But there was another network in place, he pointed out, that could convey long-distance signals from one part of Canada to another, and it belonged to CNCP Telecommunications. According to Rogers, a Cantel subscriber could make a call, which would be routed through a switching station and transferred to CNCP's satellite network, then distributed through a switch at the other end to its destination.

Meanwhile, Rogers continued to expand the capacity of his own cable installations with an eye on the day when it would carry far more than a bunch of TV programs. By fiddling with signal frequencies, for example, Rogers technicians began to relay more than one signal over a single communications line. This is called multiplexing. Because it requires fewer transmission lines, multiplexing significantly reduces the cost of constructing, servicing and maintaining a communications network.

Likewise, through a process called compression, Rogers could reduce the amount of bandwidth required to transmit TV pictures and thus increase channel capacity by four times or more. Channel capacity had restricted Rogers's ability to offer the same program at a number of different times.

Even if you broadcast *Death Wish, Dead Zone, Death in Venice, Die Hard* and *Die, Die My Darling* at 9:00 in the evening, the subscriber is a captive of time. He has to put the kids to bed, find the bread and the mayonnaise, make a baloney sandwich, put it on a plate, fill a bowl full of potato chips, grab a Diet Pepsi from the fridge and be in front of the TV by 8:55 p.m. or he'll miss the opening frames of the movie. If you could offer one or all the movies at several times throughout the evening, you could give subscribers a choice. They could take their time, would feel more in control of their lives, and would pay you buckets of money for the service. As Rogers observed, "When you mention 150 channels or 1,000 channels, it's not to have a thousand different services on, it's to provide an opportunity for the repetition of many of the existing ones, so that at a time of your choosing you can see it."

In 1992, Rogers's First Choice pay-TV service began using compression to deliver movies on four channels. Subscribers could now pick from four movie titles at a particular point in an evening and select from a total of twelve movie offerings every night. Eventually, cable and satellite services will offer sixty to a hundred channels of movies, sports and the latest installment of *How to Nail Your Cat to the Floorboards*. They will be able to record movies for a price on a smart-TV equipped with a compact disk. Compression will also reduce the capacity required to broadcast a signal via satellite, whose owners now charge $1,000 an hour or more to transmit programming.

Satellites have been floating overhead for more than thirty years, since *Telstar* was launched on the night of July 10, 1962. *Telstar* was used to transmit the first live transatlantic television broadcasts between the United States and Britain.

In 1965, the Communications Satellite Corporation (COMSAT) launched *Early Bird*, the world's first commercial satellite. In addition to TV programs, *Early Bird* relayed telephone messages between Europe and the United States. In Canada, Telesat, an organization jointly operated by the federal government, the phone companies and CNCP, launched the first *Anik* satellite in 1973. (*Anik* means brother in the Inuit language.) It hovered in space like a little brother, 38,600 kilometres above the earth, flying at about 11,000 kilometres per hour so it would remain over the same spot as we hurtle together through the cosmos.

By 1994, five generations of *Anik* satellites had been launched, each more powerful than its predecessor. Until January of that year, nothing had ever disturbed the satellites' steady transmission of signals. They were the Cleaver family of outer space – Ward, June, Wally and Theodore – solid, respectable, consistent and predictable. Then on January 20, a barrage of space particles caused the failure of a solar-powered stabilizing wheel, and *Anik E1* moved out of its standard alignment, disrupting the signals of a total of seventy conventional and cable-TV broadcasters, radio-program syndicators and fifteen voice data businesses, as well as the *Globe and Mail*, which uses *Anik* to transmit the breathless prose of its writers to printers and on to lucky consumers throughout Canada. Disbelieving Telesat personnel worked for eight hours to correct the problem. But no sooner had they corrected that failure than *Anik E2* suffered a circuit failure. It too spun aimlessly in its orbit, scattering radio and TV signals towards Mars and Jupiter instead of Nepean, Ontario. It was a bad night for satellite service, and it cost Telesat more than $500,000 to reimburse its clients for lost service, even though Telesat rented space temporarily on a U.S. satellite and restored most of its *Anik* service in less than eight hours.

On good days, which occur most of the time, *Anik* and similar vehicles transmit their signals back to earth over a fairly wide area. Cable companies, bar owners and other dubious characters equipped with satellite dishes the size of their garage have no trouble retrieving these signals. By 1991, in fact, there were an estimated 300,000 of these jumbo dishes in Canada, even though Canadian law prohibited private satellite reception by individuals. Recently, more powerful satellites have been launched that can retransmit signals back to earth with more precision to a smaller dish no bigger than Wendell Clark's hockey helmet, and the CRTC has relaxed its restrictions on private ownership.

Operators of direct-to-home satellite companies such as Power DirecTV, which transmits via a U.S. satellite, and Expressvu, which bounces its signal off *Anik E2*, claim that their services will cost sub-scribers $60 a year less than cable. Charging about $10 a month for extended basic service and $12 a month to rent the satellite receiver, Expressvu anticipates signing up 100,000 subscribers in its first year and covering the cost of customer hardware in four. Meanwhile, three Canadian services already on DirecTV in the United States stand to make $400 million in subscriber fees alone in the next eight years, according to the hype that preceded the company's application to operate in Canada. If Canadian programmers captured just 1 per cent of the total U.S. pay-per-view audience, Canadian producers would earn $10 million. Film and TV producers, meanwhile, have predicted their share will be 4 per cent.

Even before the satellite broadcasting companies began manoeu-vring for approval of their services from Ottawa, Rogers invested in a satellite company. Cancom, which Rogers owns 20 per cent of, dis-tributes nine Canadian and American television signals by satellite to more than two thousand cable systems across Canada. Most systems are in small, remote communities that depend on satellites for their mainstream broadcasting signals. In turn, Cancom owns 19 per cent of Expressvu. Other shareholders include BCE Inc. with 33 per cent, WIC Western International Communications Ltd. with 14 per cent, and Tee-Com Electronics Inc., which manufactures the antennas and set-top boxes to receive Expressvu's transmissions, with 33 per cent. "It's a portfolio investment," explained Colin Watson, "and Cancom has a future irrespective of DTH broadcasting."

Rogers has also pledged to spend several million dollars to help small cable companies compete with direct broadcast satellite ser-vices. "The Canadian cable network needs a national, satellite-fed cable facility," he said in 1994. "The industry also needs national phys-ical interconnection to allow services to flow seamlessly from coast to coast." To this end Rogers says his company will spend $8 million on "Head End in the Sky," a satellite uplink and digital video compression system. But this amount pales compared to the money needed to upgrade Canada's cable network so it can deliver digitally compressed material. According to the cable industry, that task will cost $6 billion over the decade.

As for his own company, Rogers continues to expand its pay-per-view channel capacity to twenty from four, giving subscribers the equivalent of video on demand and allowing Rogers Cablesystems to compete with direct broadcast satellite. It also plans to introduce digital video compression boxes in 1996, at a cost of $500 per box.

By 1988, Rogers was well aware that the major threat to his company came from the telephone companies. That year, he launched a $600-million, three-year spending program to link Cantel's cellular systems in Nova Scotia, Quebec, Ontario, Manitoba and Alberta into a 7,800-kilometre continuous network, the largest in the world, covering all ten provinces from sea to shining sea. He also raised about $1.63 billion from the sale of his U.S. cable holdings to build a war chest for his looming battle with the long-distance phone companies. At the time, Canadians were paying 20 per cent to 25 per cent more than Americans for long-distance service provided by Telecom Canada (now called Stentor), owned by Bell Canada and nine local phone companies. This premium was particularly critical for large companies that send voice and data via long-distance lines. In the United States, long-distance rates had fallen by about 40 per cent in five years after competition began in 1984. The number of long-distance transmissions of voice and data, especially with the rise in popularity of fax machines, rose by 46 per cent. Revenues doubled to $60 billion in 1989, while costs were down 70 per cent, thanks primarily to fibre optics.

Rogers was already providing high-speed data links to businesses through his cable network. And common standards were being developed that would make phones, computers, fax machines and cable TV compatible with the same wire. Rogers even had switching systems in place that could handle the two-way communication of large volumes of information.

Switches had traditionally distinguished the phone companies from the cable companies. By allowing information to pass in each direction along a single communication line, a switch allows for conversations rather than simple viewing or listening. For years, phone systems were switched networks; cable systems weren't. Phone companies and their research facilities continue to produce ever-faster switching systems. Bell-Northern Research, for example, has developed a concept for an inexpensive terabit switch capable of handling digital voice, data and video at trillions of bits per second. (The fastest switches currently in use handle one or two billion bits per second.) The design uses optical fibre to link numerous high-capacity switching systems into one giant switch, capable of handling the hefty demands of multimedia networks. It increases by one hundred times the speed of the fastest switches currently in use to route customer calls. Such a system could simultaneously route 40,000 video conferences or movies to customers each second, according to a report in the *Wall Street Journal*.

But the phone companies haven't been alone in trying to cram information elephants into digital thimbles. Rogers also knew that by combining switch technology with video transmission by fibre-optic cable or other means, he would end up with interactive video, and he had the resources to try it himself. Cantel operated an extensive switched network, and Rogers's installation of fibre-optic networks would expand his switching capacity even further. "The telephone companies would like us to use their fibre-optic line into the house for local calls, cable and long distance. Why can't my line be used for all the same services?"

Fibre-optic cable consists of hair-thin strands of glass that carry laser-generated pulses of light. The light waves conduct voice, data and video signals at high speeds. The cost of manufacturing and installing fibre-optic cable is now as low as the cost of copper wire. In fact, the cost of copper is rising, while the cost of fibre's raw material – sand – is not.

One fibre-optic wire can carry all voice, data and video messages to a house or office. And signals travelling over fibre-optic lines don't have to be amplified as often as analog signals over coaxial cable. Fibre-optic lines do have some drawbacks: They can be affected by impurities in the line, or by smearing of the pulses. They're also extremely difficult to splice.

Fibre-optic cable in 1989 was used primarily on trunk lines, but Rogers started to devote millions to completing fibre-optic rings around Toronto and Vancouver. Buying cable from companies like Pirelli, Corning, and Northern Telecom, Rogers and the phone companies also prepared to link fibre-optic lines directly to homes and offices, at a cost of billions of dollars. "We see great potential in PC access services, home shopping, telemetry and interactive video services," Rogers said. "And by putting programming on a computer, people can access it at any time. The future is switched more than anything else."

Although telephones, TVs and radios remain primarily analog devices, a digital world is not far off. Meanwhile, experiments with two-way video have been tried with much fanfare but little consequence since the 1970s. In 1977, the Qube two-way interactive television service was launched in Columbus, Ohio. Subscribers could tailor their cable TV service to match their viewing requirements. With a pay-per-view programming option, they could receive selected entertainment and sporting events for a fee. A keypad control linked the subscriber to a central computer. Qube subscribers could also interact with programs. After a speech by Jimmy Carter, viewers were invited to participate in an electronic survey by responding to a series of questions that appeared on the TV screen. The experiment did not expand, however, and Qube now provides subscribers with cable and pay-per-view features but not interactive capabilities.

Those were the good old days. By the 1990s, there were more reliable and accessible ways to make TV interactive, and Rogers was determined to find them. In 1994, for example, Rogers and Microsoft agreed to use the software manufacturer's Tiger software to develop an interactive TV service. With a set-top box manufactured by General Instrument, Hewlett Packard, Intel, and/or Compaq, a consumer could not only order movies on demand, but could interact in a virtual community through his TV or personal computer. "We're looking at doing for the entertainment industry what desktop publishing did for the publishing industry," said Ken Nickerson, director of technical services for Microsoft Canada. "In the near future, you can have an office with software like Softimage providing a virtual studio" comparable to one now equipped with film crews and millions of dollars in equipment. "If you have a home video of the family out on a picnic," Nickerson continued, "now you can add a few dinosaurs roaming around in the background, and you'll be able to do it with $5,000 to $10,000 worth of equipment and software.

Added Steve Pawlett, editor of *Cablecaster* magazine: "What we will see happen at this point is much narrower niches will be filled much like you see in the magazine world. Imagine your own

Canoeing in Algonquin Park show. This may very well be where we are headed, but how and when we get there isn't quite as clear."

What remains clear is a continued barrage of advertising, in one form or another, some more obtrusive than ever. Through a box installed in a consumer's home, for example, companies can issue coupons to consumers as they run their ads on TV. In a test run, McDonald's ran commercials offering two burgers for the price of one. If the viewer clicked a button on his remote control device, the machine printed the coupon. Sixty-eight per cent of viewers clicked, and more than half remembered to bring the coupon to the store (presumably a major problem with ordinary coupons clipped from newspapers). The net redemption rate was 38 per cent, more than thirty-eight times higher than the redemption rate for coupons in general, which is less than 1 per cent.

To add another link to his fibre-optic network, Rogers formed an alliance in 1989 with CNCP Telecommunications. The company had been generating profits from its telex service of $14 million a year in 1986. In only three years, as fax machines took the country by storm, CNCP was losing $7 million a year. However, the company had recently completed a digital microwave and fibre-optic network from Montreal to Vancouver, at a cost of more than $700 million. The network includes a microwave tower every 48 kilometres, from one end of the country to the other. By linking CNCP's network to his own cable system, which serves more than 87 per cent of the homes that his lines pass in Toronto, Calgary and Vancouver, Rogers could simplify long-distance billing and provide another service to compete with the phone companies.

It took until June 1992 for the CRTC to approve long-distance competition. But the alliance started immediately to use Rogers's fibre-optic rings for the local loops in its national data network, which did not require CRTC approval. By the time the CRTC gave its blessing to long-distance competition, businesses could connect directly from Rogers's fibre-optic network into CNCP's backbone network, a $381-million joint system, bypassing the local phone companies and slashing their voice and data transmission costs. By 1992, two major Canadian banks were connecting their offices in Toronto and Vancouver using the Rogers fibre-optic and satellite broadcasting system. And by 1993, Rogers had signed every major bank but one to conduct high-speed financial transactions among their data centres in Toronto, through Rogers Network Services, a division of Rogers Cablesystems. As George Harvey, president of CNCP, and later Unitel, said, "Technology makes savings probable. Competition makes savings possible. Customers' choice makes savings inevitable."

When it opened for business on October 19, 1992, the Rogers-CNCP alliance, now called Unitel, still had to assign each customer a seven-digit access code and a ten-digit personal identification number. Although a customer could program the codes into the phone's speed-dial mechanism, and although businesses could use a special black box to dial the codes automatically, it was an unwieldy system. To remember a Unitel seventeen-digit number was like reciting "The Night Before Christmas" every time you wanted to make a phone call.

The phone companies were delighted. To add insult to injury, they tied up Unitel in the courts to prevent them from gaining access to phone company technology. To reach a Unitel hookup, calls had to travel through a phone company's local switch or a business customer's Private Branch Exchange. Then the call travelled through Unitel's national digital network and back through the local phone company's switch at the other end. The phone company at each end charges Unitel an access fee. Between 1991 and 2006, fees and transfer payments to the phone companies will likely cost Unitel $11 billion. In the United States, local phone companies collect $25 billion a year in access fees for allowing long-distance companies to complete their calls. Rogers has complained that these fees are exorbitant.

He's been joined by other companies that have entered the long-distance market. Like airlines that sell empty seats at a discount, phone companies sell empty long-distance space to resellers. Buying long-distance services in bulk from the phone companies, resellers peddle long-distance calls at a discount to business users. Unlike the phone companies, and Unitel, resellers have low overheads, because they don't have to pay for equipment or the staff to maintain it. By 1990, even before the CRTC approved long-distance competition, resellers held 2 per cent of the phone companies' market share for private business long-distance service.

When it opened for business, Unitel aimed publicly to take 20 per cent of the long-distance market by 2002. Its share would be worth about $1.2 billion. To achieve its goal, it pledged to spend $750 million to $1.5 billion over five years to develop a truly competitive long-distance system. While consumers chatted and faxed to their hearts' content, their conversations would travel along their television cable to a central switching station and through the Unitel system.

In the meantime, Unitel had to overcome a bad reputation for quality, to which the phone companies did not hesitate to draw their customers' attention. "It wasn't too long ago that Unitel had a horrible reputation in the industry," said Judith Cole, a telecommunications consultant with DMR Group in Toronto. Salesmen didn't understand the technology, she observed, and the system frequently failed. But by 1992 most of the problems had been worked out, especially with the installation of CNCP's national fibre-optic and microwave network. In fact, now the phone alliance was at a disadvantage, because long-distance calls had to be channeled through a series of networks owned and operated by the regional phone companies.

But Unitel still had to persuade customers to switch, and most customers couldn't have cared less. "A phone is a phone is a phone," observed Mark Lowenstein, a fan of Gertrude Stein and a program manager at the Yankee Group, a market research firm. "There's no room for product innovation. But there is room for service innovation."

In this area, Unitel offered comprehensive computerized billing, which provided business customers with a variety of data organized according to the customer's requirements. If a customer wanted to see how many hours its sales staff spent on the phone, Unitel could break down the bill by times and duration of call; if it wanted to see where the staff was calling, Unitel could provide them with a breakdown of area codes and time zones. Unitel also offered virtual private networks. Operating on public phone lines, these networks allow employees at one office in Halifax to communicate with employees at another office in Moose Jaw just as they would if they were trying to contact a co-worker on another floor through a dedicated private network, but at a fraction of the cost.

Rogers already had experience offering private-line voice and data services to business customers through Rogers Network Service. Now, combining cable, cellular and phone technology, Unitel and Rogers could provide a mobile network of microwave and fibre-optic systems that would allow customers to communicate by phone, fax, personal computer or wireless technology through local and long-distance channels. "We see great opportunities to expand our products and services within this existing business," Rogers said in early 1995.

"In the next few years, Rogers Communications will be providing as many services to home computers as it will to television sets,"

Rogers continued. "A television set or personal computer connected by a modem into the cable system will deliver two-way signals for all kinds of transactions, at speeds 1,000 times faster on cable than over telephone lines."

Rogers expects to upgrade its cable network to two-way service by 1996. The system that Rogers aquired from Maclean Hunter, which did not invest as much in its technology, will take a year longer.

Meanwhile, with the introduction of digital personal communications devices, Rogers also sees a promising future for Cantel. By 2010, the industry is expected to be worth US$30 billion, with up to 45 per cent of North Americans using wireless in some form or another. Wireless phone systems, called Personal Communications Service, enable the transmission of digital signals carrying voice, data, fax, even video. PCS is expected to compete with the traditional phone market.

All this lies in the future. But already, the phone companies are invading Rogers's turf, conducting pilot projects to compete with pay-TV and offering video on demand. In Saskatchewan, SaskTel has experimented with pay-per-view fibre-optic cables to carry movies. The experiment extended to 136 home screens whose owners could call in and program their individual video requests. By calling a number on their phone, residents could request one of two dozen movies, delivered along the phone company's fibre-optic trunk lines and into the home via coaxial cable.

When SaskTel conducted some unauthorized testing of video signals over its phone lines, Regina Cable launched a $28-million program to equip the city with fibre-optic cable of its own. It also installed several thousand aptly named Impulse boxes that allow participating cable subscribers to make pay-per-view selections at the flick of a button so they can watch Hulk Hogan pummel the Beast for $4.50. It also allows the cable company to send tailored messages to its subscribers, mostly for marketing and promoting pay-per-view selections.

Farther afield, in Florida, Northern Telecom has provided a new suburban community of four thousand houses near Orlando with pay-per-view home screens, picture telephones, home-banking services and other wonders of the information age.

But in Rochester, New York, an experiment with set-top boxes that allowed video-on-demand was a disappointment. Instead of participating avidly in the experiment, subscribers often continued to rent their movies from the local video store. Rochester Telephone took the boxes back after a year.

This hasn't discouraged the phone companies from diving head first into the new age. In 1994, Canada's telephone companies, operating under the name Stentor, pledged to spend $8 billion on an initiative called Beacon to bring fibre optics within reach of 90 per cent of all Canadian telephones within a decade. An analyst called it a "pre-emptive strike against the cable companies." Through an affiliate called MediaLinx, the phone companies also plan to provide two-way video messaging, electronic full-motion shopping, distance education and dial-a-movie services, all with a simple telephone call.

When they're finished, the phone companies say they will lease space on their fibre-optic network to the cable companies. But Rogers, for one, doesn't buy their scheme. "Dumb cable companies would go for it," he said. "But in the end they'd just be bill collectors, employees of the phone companies."

According to Rogers, the phone companies don't need fibre-optic cable to handle phone calls. "For what reason would they do it, and who would pay?" he asked. "Certainly the interest alone on the installation of fibre-optics would cost more than the monthly phone service. Are they going to charge each phone subscriber that huge added cost?"

True to Rogers's forecast, as the cost of local phone service rises, phone companies have found ways to modify their existing copper wires to increase the data capacity. AT&T, for example, has developed a technology called VideoSpan-Plus that enables copper telephone wire to behave like high-speed fibre-optic cable. The technology is being tested in the United States by Bell Atlantic and in Hong Kong, South Korea, Sweden, Italy and Britain. The technology brings telephone companies one step closer to their goal of offering cable-TV services to their customers, bringing the cost down to about $500 per subscriber line, reduced from the $2,000 a month currently charged for dedicated, high-speed telephone circuits. "I wouldn't be surprised if we announce live broadcast capability for VideoSpan-Plus by year's end," said an AT&T spokesman in 1995. Without the technology, phone companies had estimated it would cost $600 billion to dig up existing copper lines and replace them with fibre-optic cable.

At the consumer end of the line, digital TVs and monitors will soon replace bulky units with flat models about three inches thick that you can hang on the wall between sessions next to your print of Picasso's *Woman with Child.* A company called U.S. Robotics has developed a fax-modem that enables PC users to talk and share data at the same time using a standard phone line. Long-distance calls can now be made over the Internet. Computer-readable audio files can deliver AM radio programming over the Internet, as well, and radios themselves will soon become digital.

As technologies emerge and converge with mind-numbing speed, it becomes apparent that technology itself isn't really all that interesting and has little to do with the ultimate outcome of the battle between Rogers and the phone companies.

In fact, technology isn't the issue. Regulation is. As Eric Manning, dean of engineering at the University of Victoria, said in an interview with the *Globe and Mail's Report on Business* magazine: "The big question is who is going to install it, who is going to own it, who is going to operate it, and who is going to make the money from it? If the phone company's got the fibre line, you don't need the cable company. And if the cable company's got it, you don't necessarily need the phone company."

CHAPTER 4

The Regulators

Riders on the Storm

You can die of boredom by just looking at pictures of
Canadian life.

LOUIS DUDEK

Ted Rogers held his first tryst with the guardians of Canada's airwaves in 1960, just three years after he met his wife. Since then, his relationship with our communications regulators has lasted as long as his marriage, left him far more bewildered and given him many more sleepless nights. Regulators have pushed him to the brink of bankruptcy with one hand and protected him with the other from the jaws of unbridled competition; raked him over the coals in public, then joined him for intimate sessions in the corporate boardroom. The superannuated bureaucrats, myopic visionaries and politically anointed barnacles of the ship of state, acting on behalf of the 26 million Canadians with radios, televisions and telephones, have walked hand in hand with Rogers since he first began his adventures in communications. They practically ignored him when he bought his first cable licence in the 1967. Then they whispered sweet nothings in his ear when he expanded his cable operations into the United States. They nudged and winked when he entered the cellular industry, acquiesced when he demanded competition in long-distance service; retreated with a far-away look in their eyes when he argued against competition in the cable industry, and lay back in their lounge chairs when he acquired Canada's most venerable magazine publisher. They've also slapped his wrist, straightened his corporate tie and thrown a few bureaucratic pots and pans in his direction when he seemed to stray from their counsel. But if Rogers ever thinks that he can't live a moment longer with his irritable regulatory spouse, he also knows that he can't live without her. "If something was bothering your wife, and you wanted to remain married, you'd bloody well fix it," Rogers once said, after the CRTC had insisted that he reduce the level of foreign ownership in his company.

Like many a marriage made on earth, Rogers's commitment to his regulatory partners has brought him moments of sheer bliss and extended periods of bittersweet anguish. And like many a marriage in Canada in the 1990s, this one ran its course long ago and is now fuelled by momentum far more than passion. Until the marriage ends, however, Rogers Communications will continue to employ two full-time executives, a vice-president of regulatory law and a vice-president of regulatory affairs, at six-figure salaries, to advise Rogers from day to day on how far he can go to get what he wants without raising the hackles of his regulatory mate.

Canada's regulators lost their innocence long before Rogers ever came to the party. The U.S. and Canadian governments first began regulating the electromagnetic spectrum in the 1920s. Initially, governments focused their attention on the technical aspects of broadcasting and communications, allocating bands of the spectrum to broadcasting organizations and licensing these organizations to use them. But the public hardly cared about the frequency used by a radio station, as long as it delivered Kukla, Fran and Ollie to their living rooms. In Canada, the government soon realized it could gain a lot more attention and associate itself with much more profound concerns if it meddled in other aspects of communications, under the dubious banner of cultural affairs. Even though no one has ever offered an adequate definition of Canadian culture, the government set out to preserve it, whatever it is, by monitoring the airwaves.

With this in mind, Canada set up its first royal commission on the future of broadcasting in 1928, and it hasn't looked back. Almost seventy years later, the government continues to stick its nose into every nook and cranny of the communications industry on behalf of the taxpayers who are too brain damaged to think for themselves but not too addled to earn enough money to pay the government to think on their behalf. With an annual budget of more than $33 million and a phalanx of 430 executives, scientists, professionals and their staff, CRTC pundits have pontificated endlessly on the future of Canadian broadcasting, while most Canadians on whose behalf they labour have had only a limited choice in the variety of television programming they receive and the media that deliver it.

Based on the findings of its public hearings in the 1930s, the Canadian government set up the Canadian Radio Broadcasting Commission to regulate broadcasting, create a publicly owned coast-to-coast network and produce programs sustained by taxpayers rather than advertisers. "The use of the air that lies over the soil or land of Canada is a natural resource over which we have complete jurisdiction," said Prime Minister R. B. Bennett. He explained that the government should control the airwaves "for development for the use of the people."

Within four years, it became obvious that the only people benefiting from the commission were the well-paid commissioners. So the new government of Mackenzie King assigned the commission's responsibilities, and a few more to boot, to the newly minted Canadian Broadcasting Corporation. For the next twenty-six years, the CBC rolled merrily along, regulating all broadcasting in Canada while stifling potential competition from private operations.

In 1958, Canada assigned the CBC's regulatory powers to the Board of Broadcast Governors, leaving the CBC to hire managers, pad its budget and focus on its own organizational navel. Private broadcasters had sought their own network to compete with the CBC for decades. On behalf of the government, the Board of Broadcast Governors finally complied in 1961, when it licensed CTV, the country's first private television network. Despite the board's apparent arm's-length relationship with the government, however, accusations flew fast and furious about political interference with programming and licensing. In gaining favour from broadcasting authorities, it seemed, who you knew had become as important as what you knew. Once the government noticed the problem, it responded as only a Canadian government can: It called for a royal commission.

After another tedious round of hearings, the Board of Broadcast Governors was replaced by the Canadian Radio-Television Commission on April 1, 1968. This was no April Fools' joke. Pierre Trudeau, the newly elected prime minister, seriously intended that the CRTC should act on behalf of his more feeble-minded citizens – a category to which all Canadians but he belonged – in regulating programming and enforcing standards that would indoctrinate us all in the subtleties of Canadian culture. In pursuit of the true meaning of life as a Canadian, the CRTC's five full-time and ten part-time members set about mandating Canadian content and looking askance at programming that smacked of philistinism or non-Canadian values like humour, or sound that didn't seem to emanate from the bowels of an aluminum trash can. They were especially wary of programming that came from the odious hell-hole south of our border – the longest undefended border in the world, lest we forget, and wide open to any squinty-eyed soul-sucking seditious idea that can strap on a six-gun and gallop down an airwave into the true north strong and free. The CRTC also ordered cable companies to originate local public-service programming so that all of us could gather round on Tuesday evenings at seven and watch the mayor of East York discuss bi-weekly garbage collection.

For more than two decades, the CRTC has endeavoured "to safeguard, enrich and strengthen the cultural, political, social and economic fabric of Canada," a simple task that might challenge the twelve apostles but not our fifteen politically appointed commissioners. Through the commission's efforts to truss this evasive mandate to a board and skewer it with rules to keep it from wriggling, the remaining 26 million people in Canada who don't sit on the CRTC have been privileged to curl up in front of their TV sets to watch *Juliette* and *Front Page Challenge*, while our American neighbours degenerate into slavering psychopaths and adulterous faith healers under the nefarious influence of *The Boob Channel* and *Gardening on Steroids*.

As the technology of communications has evolved, so has the CRTC's power, along with its name. From allotting positions on the electromagnetic spectrum in the 1930s, the CRTC now holds sway over everything from the number of baseball games you can watch on an American sports channel carried on your Canadian cable network (essentially, none, unless you pay dearly), to the type of equipment you can attach to your phone line, to the amount of money a supplier can charge to provide you with its limited, over-regulated service.

Now called the Canadian Radio-television and Telecommunications Commission, the CRTC has influence that extends from cable TV and broadcasting to cellular and long distance. But as wide as it casts its regulatory net and as often as it adds words to its name, the CRTC cannot keep pace with the rapidly evolving technology from which it hopes to protect us all.

Like the pope dispensing favours and collecting overdue accounts in God's name, the CRTC giveth and the CRTC taketh away in the name of the government in power. Unlike God, however, the government occasionally shows up on the CRTC's doorstep if it doesn't approve of the commission's work. In 1994, for example, the CRTC approved an application from Expressvu to deliver satellite TV service to Canadian subscribers. The CRTC rejected a similar application from Power Direct, ostensibly because it would bounce its TV signals off an American satellite rather than a Canadian satellite. The distinction was irrelevant to all but the owners of the satellites and the commissioners themselves. In fact, the government had first studied the issue of satellite TV in the early 1980s. By 1995, as their government and its appointed commissioners still dithered over the issue, about 30,000 Canadian consumers had already purchased equipment to receive satellite signals, no matter how they arrived at their Canadian homes. After much gnashing of political teeth, the government finally ordered the commission to reassess the applications and open satellite service to competition.

Because it enjoys a much higher profile than similar organizations like the Bureau of Pension Advocates, the CRTC provides an easy target for politicians hoping to draw attention to themselves. In 1990, the CRTC allowed cable companies to increase their rates for basic cable service to cover the cost of new equipment. In return, the cable companies would have to reduce their basic rates by the same amount five years later, after they'd covered their costs. In 1993, however, the CRTC modified its ruling. Cable companies could now keep the increased rate in place for basic cable service, if they agreed to contribute half the increase on a monthly basis to a fund for producing Canadian television programs. For Rogers, this meant a contribution of about 80¢ a month per subscriber, or $23 million a year.

This was too much for Liberal MP Dan McTeague of Whitby, Ontario. In March 1995, McTeague drew himself up to his full political height, peered over the back bench and sputtered that the CRTC planned to gouge Canadians with hidden cable taxes of $1.1 billion over the next five years. "The CRTC may think they're arm's length from Parliament and the people," stated McTeague, "but they've demonstrated they're not at arm's length from the cable companies.

"The CRTC has been playing fast and loose with the money of Canadian cable subscribers for some time and now they are doing the same with the truth," McTeague continued. "The money grab by the CRTC and the cable TV companies is indeed hidden because Canadians do not know that their rate reductions have been hijacked by revised CRTC regulations." In other words, Canadians would be upset if they ever found out that the CRTC had not reduced their cable rate after they'd toiled and sweat for five years to eke out an extra 80¢ a month.

Regardless of McTeague's accusations, the CRTC derives its authority from a skein of legislation that keeps Rogers and his regulatory advisers in their offices long after most people have gone home to watch *NYPD Blue*. To administer policies affecting broadcasting and cable, the commission is guided by Canada's Broadcasting Act. The act falls under the purview of Heritage Canada, the Redi-Whip of political departments, which is also responsible, as much as anyone can be, for the CBC. Industry Canada, a more durable bureaucracy, in name at least, sets policy under the Telecommunications Act, passed in 1993. Under the auspices of this legislation, the CRTC regulates activities in cellular and long distance.

The Broadcasting Act lays down rules governing the content of programming delivered by Canada's radio and TV stations. Under the act, the CRTC can restrict entry to broadcasting and require a station to carry a predetermined amount of Canadian programming. Of the mindless pap carried by a Canadian rock-and-roll station, for example, 30 per cent must be Canadian. Of TV programming, 60 per cent must be Canadian. AM radio stations are regulated according to format and commercial time; FM stations according to format, although the rules have been relaxed in this area. As with all pointless attempts to define ourselves, the definition of Canadian content is elusive. You'll know it when you see it, and if you don't, the CRTC will tell you.

Because of the Broadcasting Act, Rogers Cablesystems cannot carry a U.S. cable service that competes with Canadian services. As a result, Canadians can watch TSN but not ESPN; Much Music but not MTV; the Country Channel but not the Nashville Network. The CRTC also must approve packaging of services and the types of signals that cable companies can carry. And finally, when all the equations are complete, a cable company must carry more Canadian than foreign services.

A twenty-five-year-old cabinet directive, also passed during the long-haired Trudeau days, restricts foreign ownership of Canadian radio stations to 20 per cent and gives the CRTC the power to define the meaning of "foreign." So far, foreign means not Canadian; Canadian means not foreign. Anne Murray is a great Canadian, for example; Bryan Adams isn't; Neil Young is sometimes, but not always; k.d. lang could be Canadian if only she would try; your neighbour next door who socks back six Red Stripes and starts playing his steel drums at

4 a.m. would be Canadian if he recorded a song in a Hamilton studio, but will remain a mere landed immigrant as long as he just sits around shaking his dreadlocks, inhaling doobies and pounding on those oil cans; and so on.

As with all regulatory bodies that work at arm's length from the government that authorizes their activities, the CRTC behaves in strange and unpredictable ways, based on reasoning to which none of us has access. For the last few years, the CRTC was believed to be in a sympathetic mood to cable TV in general and to Rogers in particular. After years of responding like the Ice Queen to overtures from Rogers and other cable companies, the commission now seems to have swallowed Rogers's arguments that a strong cable industry can provide Canadians with an appropriate dose of Canadian culture through their television sets. In reaching this decision, which isn't really a decision, the regulators have allowed cable companies to pass along some, but not all, the costs involved in upgrading their cable systems and equipment. They have also prevented telephone companies from delivering TV programming over their phone lines. In the process, the commission seems to want cable companies to compete with satellite services while growing big and strong, like the phone companies. Meanwhile, no one knows when the rules, which aren't really rules, may change again.

Regardless of the prevailing mood of the CRTC, cable companies like Rogers still have to abide by the Broadcasting Act. In matters involving basic service rates, content and geographical service areas, for example, companies must approach the CRTC on bended knee and ask for permission to proceed. The CRTC may deign to grant a rate increase, for example, if a cable company can show convincingly that it will have to incur great expenses to add a new channel to its system and that the customer should carry the cost, or if it can demonstrate a pressing need for new equipment or systems, or if it simply needs the dough. Rogers can charge up to 5¢ per subscriber per month for a new channel. His company can increase cable rates by up to 3 per cent to pay for capital expenditures. Or, if his piggybank is empty, he can raise rates according to a convoluted formula based on return on average net fixed assets, before interest and depreciation, world without end, amen.

While Rogers has to jump through regulatory hoops to raise basic cable rates, other aspects of the industry remain entirely unregulated, except by the power of the marketplace. Rogers can charge a subscriber as much as he wants to rent a converter, for example. He can also jack up the cost of pay-TV without ruffling the feathers of the CRTC. In these areas, the rules of the marketplace prevail, and they do not respond so predictably to Rogers's overtures.

The marketplace revolted, for example, in 1994, when Rogers and every other cable operator in the country raised the monthly fee for its extended cable service to cover the cost of some new specialty channels. The CRTC had approved the seven new channels, including the Life Network and the Women's Network, several months earlier. As they had done many times before, the cable companies sent a notice to subscribers announcing the new channels. Subscribers could receive all of them, at an additional cost, without doing anything at all. If subscribers didn't want to receive the new channels, all they had to do was return a card to the cable company.

Critics saw this as an egregious and underhanded tactic to squeeze more money out of downtrodden cable subscribers, who might end up paying for the new service whether they wanted it or not, just because they didn't return their cards to the money-grubbing scheming cable companies.

While Rogers's competitors sat back and blew on the flames, a blazing brouhaha ensued, out of all proportion to the misdeed. At one point Rogers Cablesystems was receiving ten thousand calls a week; and four thousand of its 2.9 million subscribers discontinued their optional service.

Rogers took the pounding gracefully, and tried to turn it to his own advantage by adjusting the company's policy on rates. Though he never said as much, the fault lay with the CRTC. Under its rules, cable companies must offer one Canadian service for every American service they carry in a package. To comply with the guidelines, cable subscribers could take all seven new services or none; they couldn't pick one and reject another without upsetting the regulatory balance of foreign and domestic services. Even the Friends of Canadian Broadcasting, who usually regard Rogers as they would a pimp at a Sunday-school picnic, sided with his company this time. "We're not blaming Rogers," said the Friends, "they're just behaving the way any company would." Of course, the Friends' sympathy was not unqualified. "We're blaming the CRTC, because the commission allows these guys to charge whatever they like for all the services beyond the basic package."

When the flames died and the embers cooled, the victims of the imbroglio turned out to be the seven new specialty channels. Cable companies had guaranteed them between 25¢ and 35¢ a month per subscriber, based on full penetration of the cable companies' markets. Because of the public furor, far fewer subscribers signed up for the channels. The cable companies renegotiated a lower fee, and the new channels received far less money than they'd anticipated.

Nor does the marketplace prevail when Rogers wants to add another fiefdom to his empire. To acquire a cable system operated by another company, Rogers needs permission from the CRTC. In such cases, Rogers and his entourage usually don their sober suits and troop down to Ottawa to stage a well-rehearsed and suitably restrained performance. This never fails to influence the commissioners, attract media attention and give Rogers's cable operations a much higher profile than the two thousand other cable systems operating in the country.

With the help of his team of advisers, Rogers has navigated the sea of red tape more successfully than any other entrepreneur in Canada. He regards his presentations to the CRTC as command performances, and he approaches them much as a conductor would approach a performance of Beethoven's Ninth. He prepares every detail of his appearances, down to the last clause and the final digital pause. His approach today remains much as it was twenty years ago. His thoroughness, combined with his own zeal for the topic at hand, never fails to impress the CRTC. "It was quite clear to any commissioner that Rogers had done his homework, gone through a plain rehearsal and a full-dress rehearsal," observed former commissioner Roy Faibish of a Rogers road show in 1980. "The whole presentation was beautifully orchestrated, and therefore became compelling. Phil Lind, who looked after regulatory issues, was very polished, very likable, culturally caring, with an Ipana toothpaste smile. John Graham, Ted's stepfather, was a kind of helmsman, the Rock of Gibraltar. He wouldn't say very much, but he would closely watch Ted. And Rogers, who at times can take off on a very short runway, if he got a touch off point, Mr. Graham would send him a series of nonaudible signals and rein Ted in. Colin Watson, who probably knew as much about cable as anyone in Canada, would lend gravitas to the Rogers team."

Even then, after more than a decade in the industry and as his company's spending was rising incrementally, from $10 million a year in 1979 to $32 million in 1980, Rogers played the role of humble supplicant in his approach to the CRTC. For more than twenty years after he bought his first radio station, he was regarded as a brash upstart risking his corporate skin to enter a den of hardened big shots protected by teams of well-heeled lawyers and bureaucrats. In 1979, when he appeared before the CRTC to seek approval for his acquisition of Canadian Cablesystems Ltd., industry representatives regarded Rogers and his crew of advisers as "little guys with lots of heart" engaged in a "a classic fight with the corporate lawyer types."

But his company suddenly grew too big for the role. By 1994, when Rogers earned his six-millionth frequent-flyer point travelling to Ottawa yet again, this time to seek permission to buy Maclean Hunter, he was no longer regarded as one of the little guys. Now he occupied a loftier position in the communications hierarchy, more like Beelzebub to the CRTC's archangel. Now, when he pronounced on issues like the sad state of the CBC, people listened and asked for his solutions. When he elaborated on his plans to become a Second Force in Canadian telecommunications, people believed him. He had even been heralded, not entirely in jest, as a potential chairman of the CRTC and president of the CBC. In 1994, he himself was surrounded by an entourage of fourteen lawyers, advisers and executive hod carriers as he sought the commission's blessing on more than sixty separate applications arising from the takeover. The players had changed somewhat, but the approach was the same. Phil Lind was still a member of the team. But now the chairman was Gar Emerson, who had replaced John Graham a couple of years earlier. Graham Savage had become the chief financial officer, and Ron Osborne, still president of Maclean Hunter, was there to lend credibility and keep a straight face as Rogers himself explained why the acquisition was good for Canada, good for the company, good for Maclean Hunter and good for cable subscribers, magazine readers and just plain folks from coast to coast.

Rogers and his team left Ottawa confident that they had given their best performance. In the following weeks, while the commissioners ruminated, he added one final flourish to his performance. In a gesture that was breathtaking in its audacity, Rogers appointed Pierre Juneau as interim trustee of Maclean Hunter. The regulators could hardly doubt Rogers's sincerity when one of their own was representing him. They complied with his wishes almost without a peep.

For Rogers, this was only one step in the regulatory dance. After two-stepping to the Broadcasting Act, he can pick up the tempo as he boogies in time to other legislative rhythms. The Radio Communication Act, for example, also under the purview of Heritage Canada, authorizes the CRTC to allocate radio spectrum to cellular operators like Rogers. Meanwhile, the Telecommunications Act gives the CRTC authority to approve cellular fees. But to obtain a cellular licence, Rogers must sashay down the street, allemande right at the stop sign and slide, slide, slide a few hundred metres to Industry Canada. Recently, a court case allowed the limited reselling of cellular services, similar to the reselling of long-distance services. This would bring added competition to the industry, currently the exclusive preserve of Rogers and Bell Mobility. Further complications will arise when Canadians start to subscribe to personal cordless telephones. Among other concerns, the band allocated by the CRTC to these devices does not correspond to the band allocated by U.S. authorities, which prevents continent-wide service.

As for long-distance service, it was regulated until 1993 under the Railway Act. Since tracks and telephone wires all ran in similar directions, the government presumably figured they could all be regulated under the same legislation, along with spaghetti, pinstripes and garden hoses. In 1993, however, some bright star in Ottawa's bureaucratic firmament realized that telephones and railways no longer shared the same roadbed, telephone poles were no longer made of wood and, if you wanted to send a message from Moncton to Moose Jaw, you could usually do it via telephone without sitting in the bar car with a miner from Temiskaming. So the Telecommunications Act came into being. The new act gives the CRTC authority to approve rates and define technical standards associated with long-distance communications. The CRTC can order a phone company to provide service, for example, even if the company can't make any money in the process. The CRTC can also order phone companies to let a competitor plug its service in to their lines. For this privilege, the competitor must pay a fee. But phone companies aren't required to explain publicly how they calculate the fee. The CRTC is supposed to keep a close but private eye on the phone companies, to make sure that the fees are fair. Under this kooky set-up, Unitel pays more than 50 per cent of its revenues to the phone companies for access to their local facilities. In the United States, long-distance competitors pay less than 47 per cent of their revenues, a significant difference when revenues amount to half a billion dollars a year.

In July 1994, the CRTC ordered equal access to long-distance services by subscribers of phone companies and their competitors alike. After almost two years, this order finally made it possible for Unitel subscribers to call their mothers in Vancouver without pressing enough buttons to launch a nuclear missile. Now, just like an ordinary phone subscriber, they can dial eleven numbers and start talking while the meter ticks.

Whether it deals with television or long distance, regulation is a step removed from the political process, and many decisions are made in private. In addition to affecting what we watch on television and how much it costs to phone Squamish from Verdun, the CRTC's decisions affect the ability of companies like Rogers to sell shares, raise capital, hire employees and make a profit. But if regulations have distorted the complexion of Canada's telecommunications industry, Rogers doesn't seem to mind. After all, the same regulations have also enabled Rogers to succeed beyond his wildest dreams. "I like regulated, capital-intensive, engineering-oriented service and programming businesses," he said in 1988. After twenty-eight years of paying almost quarterly visits to Ottawa, he added, "We know about regulation."

Rogers especially likes the clause in the Broadcasting Act, reinforced by the CRTC's endless bombast since 1969, that has prevented telephone companies from owning cable operations and broadcasting licences. Like the vandals at the gates, the telephone companies would dearly love to amend this regulation so they could join Rogers and the other cable operators in raping and pillaging their way through the communications landscape.

One way or the other, however, Rogers and his team still have to dance to the regulatory polka, enduring endless questioning by commissioners and lawyers, and providing answers without divulging information. Sometimes the questions come from hostile quarters. ("Your real motive here is to keep power in the cable business, isn't it?" asked a lawyer for B.C. Tel in 1991, when Rogers was presenting Unitel's case for long-distance competition.) Some of the comments he has to field are just plain dumb. (CRTC commissioner Keith Spicer asked Rogers if the acquisition of Maclean Hunter was a good way of competing with the telephone companies. "How does this help your competitive strategy?" Spicer asked Rogers. "Is it really smart to charge seven or eight times what the phone companies charge for a second outlet? It enrages people," he added.) After a day of absorbing such drivel, Rogers said, "It's like going for X-rays of every part of your brain."

To his critics, Rogers's ability to deal with regulators qualifies his impressive accomplishments. Operating in a regulated environment, they say, he has gained an advantage not shared by other entrepreneurs, who have to survive in an unregulated marketplace, especially if they never attended law school and never learned to think in sentences as long as this one. "Rogers is a regulatory entrepreneur," sneers Professor Hudson Janisch. "All of his phenomenal successes have been in monopoly regulated markets, or largely in monopoly regulated markets. Rogers has been extremely adroit at manipulating the regulatory process to his advantage."

Such comments lose some of their sting, though, when you realize that there are two thousand cable operations in Canada and that none has grown into a business similar to Rogers, regardless of the regulatory protection they enjoy. Still there's no question that Rogers is adept at telling regulators what they want to hear in the way they want to hear it. Just as some people are better than others at job interviews, Rogers can gabble with the best of them in a regulatory hearing, placing himself a notch above his fellow cable hogs. "There's a parasitic element to the cable business, given that the cable firms don't pay for most of the signals they carry," said John Meisel, chairman of the CRTC from 1980 to 1983. "But in Rogers's case, you couldn't help but be impressed with the intensity of his arguments."

Rogers comes by his abilities to dance the fandango of regulation through upbringing and environment. Whether it involves communications, banking or health care, the regulated world presents an ideal forum for a private-school boy to apply his political and diplomatic skills. Issues are defined, rules are laid down and agreed to, and decorum is observed at all times. By showing up promptly in the right suit, stringing a few sentences together without stuttering, swearing or sweating on the podium, fielding dumb questions with good humour, doing your homework, resisting the urge to seduce the biology teacher's wife and avoiding embarrassing faux pas that might cast the institution in a bad light, you can go a long way with a little effort at UCC or the CRTC. With a lot of effort, the sky's the limit.

Unfortunately for Rogers and the CRTC, the regulatory framework within which Rogers operates has become increasingly irrelevant. Even our Friends of Canadian Broadcasting say the days of regulation may be numbered. "The whole regulatory apparatus for administering and controlling television is based on a decades-old relationship between viewers, broadcasters, advertisers and regulators," writes Friend David Ellis. "Given these changes to traditional television, we have to question whether regulations created for an industry dominated by a handful of conventional networks are appropriate in an era of narrowcasting, digital compression, pay-per-view and direct broadcast satellites."

In the view of the Friends, new technologies have "undermined" traditional regulatory frameworks. "Technology is not merely unpredictable in its effects; it can also have effects that run entirely against the grain of what policy-makers intend." While the Friends would prefer a world in which everything from cable TV to VCRs is regulated, they acknowledge that "new technologies will make it increasingly difficult for regulators to put their principles into practice and to enforce the rules."

That doesn't absolve techno-entrepreneurs like Rogers from responsibility for insinuating their dastardly technologies into Canadian lives. The Friends believe that the cable industry has "provoked audience fragmentation" and "challenged regulation" like a gang of naughty schoolboys. Canada's regulators have responded by making all of them sit in the corner with their faces to the wall and their hands on their heads, reciting Rudyard Kipling poems as the regulators devise further controls on content and distribution. As a result of the

CRTC's government-sanctioned meddling, Canadian viewers remain unsullied by foreign cultural guano while receiving fewer than half the available programs on their cable or satellite systems.

But like the little boy with his finger in the dike, the CRTC, Canadian politicians and our telecommunications giants realize that they cannot resist forever the forces of global communications in the name of cultural sovereignty. Whether they enter Canada through our PCs attached to phone lines, through TVs attached to cable networks, through satellite transmissions beamed at $1,000 receivers on the bookshelf or through the hot-water tap in our bathroom, waves of programming, interactive video, movies on demand, home shopping and other flotsam from the high seas of the information world will soon deluge Canada. As the waves crash over Canada's electronic shores, not just from New York and Los Angeles but from London, Paris, Sydney, Hong Kong and Nairobi, Rogers will no doubt stand at the helm of his communications cruise ship while the regulators hand out the life jackets.

In the United States, meanwhile, regulators have encouraged competition among systems and broadcasters alike. While Canadians watch CRTC-mandated fare such as Much Music, Americans can choose from movie channels, cartoon networks, history programs, comedy shows, travel TV and shows devoted to naked women playing volleyball on a beach. In the process, the form of the industry has evolved along with the content and the technology to deliver it. Mimi Dawson, a former FCC commissioner, speculated in an interview in 1987 that the telephone companies may dominate the distribution of information, with access to every home in the United States. TV and cable companies would become mere suppliers of programming for the new system. Such a system has been launched on an experimental basis in Ceritos, California, where a telephone-distributed information system provides subscribers with programming from cable, television and two-way interactive companies. Telephone companies may also generate their own entertainment programming, beaming it through their own network and via satellite to consumers across the country.

In Canada, the phone companies have moved a tiny step closer to the future by pledging to spend $8 billion on a fibre-optic network that will reach every home in the country. Like radios, televisions and personal computers, these networks will evolve gradually, extending the reach of fibre-optic lines to more homes and businesses in more cities while relying on conventional technologies in the meantime. None of this will happen overnight, especially if Canada's regulators don't go along with the plan. For Rogers, the longer it takes, the better. Without regulations to impede them, the telephone companies could begin competing with Rogers much more quickly.

In the United States, the idea of dismantling the FCC, the telecommunications regulatory behemoth, is gaining momentum. Peter Huber, a U.S. lawyer and senior fellow at the Manhattan Institute, recently suggested that as cable and telephone companies invade each other's geographical and business territories, the FCC should begin to wither away. "Competitive markets do not require constant monitoring," he said. In five to ten years, the FCC's work may be largely devoted to handling disputes over electromagnetic spectrum. Under these circumstances, doing away with the agency altogether becomes thinkable, Huber says. Meanwhile, regulations remain sadly behind the times, in the United States and presumably in Canada, as well. "Congress has been extremely slow in bringing telecommunications law up to date with technology," said Andrew Kessler, an investment banker in San Francisco. "We're still living in a regulatory environment that dates back to the 1930s."

The least likely but most effective solution to the distortions caused by regulations is to abolish them. Without regulators to distort the impact of competition, market forces would prevail. As Esther Dyson, a U.S. editor and writer on technology, observed recently, "A revolution will occur [in the telecommunications industry], but it will occur in the marketplace, not because of government action."

As long as regulations prevail, however, they will affect Rogers's future, particularly his profits. In 1970, regulations almost put him out of business when the CRTC decided in its wisdom that broadcasters could not own cable companies. Rogers had set up his cable company in partnership with the Bassetts, with whom he had also started CFTO four years earlier. The Bassetts had no problem buying Rogers's share of their TV station; but it took Rogers almost eighteen months of returning pop bottles, holding bake sales and rolling pennies to raise the several million dollars he needed to buy out the Bassetts from his cable operation. If it hadn't been for their forbearance, Rogers could easily have gone under.

Regulations have also made it far more profitable to run a cable operation in the United States than in Canada; they've restricted Rogers's ability to raise capital through non-Canadian sources; and they've made it almost three times as expensive to make a long-distance call in Canada than in the United States. Regulators control the placement of satellites in orbit and the amount of information that each satellite can accommodate. Unable to launch more satellites, designers have turned their attention to expanding the capacity of existing vessels and reducing their size. That hits cable companies like Rogers right in the pocketbook, because every change in a satellite's technical specifications requires the cable companies who use it to reconfigure their antennae and other equipment, at considerable expense.

If Canadian politicians ever decide to dismantle the CRTC, they will not pursue their decision overnight, or even in our lifetime. That's because few voters in Canada understand the technology involved, or appreciate the role of the regulators in controlling it, or really care much about the issue. If technology does have an impact on our culture, they would prefer to leave it to the government to deal with. A Gallup survey indicated in 1995 that almost 70 per cent of Canadians had heard about the information highway, and 62 per cent regarded it as a threat to Canada's cultural identity. They said they wanted the federal government to assume responsibility for protecting that identity.

In such a marketplace, no wonder Ted Rogers is still smiling.

CHAPTER 5

The Competition

Rogers, the Underdog

Money can't buy friends, but it can get you a better class
of enemy.

SPIKE MILLIGAN

For some of us, competition means a weekly tussle with a few over-weight ex-jocks on a hockey rink. For Rogers, it now means wrestling with elephants in a ring lined with regulatory whoopie cushions and carpeted in technological bananas.

When he first set foot in the telecommunications ring, Rogers had to contend with a few radio stations, but none was so significant that he couldn't handle the competition. When he entered the cable industry a few years later, he faced more formidable competitors like Maclean Hunter. But in the late 1960s, Maclean Hunter was a mere shadow of its current incarnation, known more as the publisher of reliable magazines and newspapers than as a technological adventurer. For many years, its cable operations hardly merited attention in *Newsweekly*, the company's employee newsletter, and MH Cable operated from offices near Toronto's Pearson Airport, next to Highway 401, far from the more sedate accommodations of the publishing company, where attendants refilled the shoeshine boxes in the men's washrooms every day so that Maclean Hunter's ad salesmen could keep their toecaps shiny. Cable had yet to become a major source of revenue for any of Rogers's competitors, and big companies used their second-string players to compete against him. In the United States, Rogers successfully went head to head against Knight-Ridder, Dow Jones, UA-Columbia and other shrewd and well-financed competitors, but again, the competition wasn't fierce, and it took place in a confined, regulated context in which Rogers and his disciples could excel.

For years, Rogers competed against his bigger rivals in the cable industry as the underdog, an upstart grappling with well-heeled fat cats, and he used this sympathetic image to best advantage. But now that Rogers has paid more than $2 billion to take control of Canada's largest magazine publishing company and openly declared war on the national telephone system, the image no longer fits. With revenues of $2.25 billion a year and more than 13,000 employees, Rogers Communications Inc. ranks among the top sixty-five companies in Canada. The public now regards Rogers as a tycoon and expects him to live up to his image. He can no longer play the role of underdog in public while privately revelling in his wealth during intimate moments at Lyford Cay playing leaners with American Express platinum cards against the front wall of his waterfront estate with a gang of his fellow UCC Old Boys. As he learned from the brouhaha that erupted in 1994 when Rogers Cablesystems tried to raise its service charges, the Canadian public has little sympathy for the rich, even if they've earned their money the hard way.

Ironically, as Rogers grapples with his image in Canada as a plutocrat, he remains a significant underdog in the global telecommunications industry in which he and his Canadian competitors must operate. In fact, when the publishing and other assets unrelated to cable are eliminated from the balance sheet, Rogers Communications hardly ranks as a player in the international arena. Even Bell Canada Enterprises, Canada's second-largest company, with revenues of almost $22 billion and 116,000 employees, is concerned about its weakness relative to its international telecommunications competitors.

Restricted by Canadian regulations, BCE cannot yet compete directly with Rogers in cable, which remains his primary business. Rogers would like to keep it that way. Like a little boy poking a stick through a fence at a tethered bull, he continually taunts BCE and the telephone companies while insisting that the rope and the fence remain in place. "BCE's profits after taxes, write-offs and real estate disasters are $1.3 billion," he chided in 1994, "and they're still laying off people. I'm not sure people making a billion-three per year need much help competing. They want to compete just long enough to bankrupt the competitors, and then it'll be just them.

"What they really want is a restoration of their complete monopoly. They would like to restore a monopoly in video back under their wing so that you'd have the mother of all monopolies. Then this fabulous information highway that everyone's talking about would be totally under the auspices of one provider, not two, not dozens. We think this is terribly wrong in terms of public interest. Innovation will slow down and rates will go up. If the telephone companies get their hands on it, by themselves, we can kiss this wonderful information highway goodbye."

As Rogers jabs and pokes at BCE, much larger competitors lurk in the shadows that could crush BCE and Rogers alike with a casual swing of their tails. Regulators realize they have little time left to prevent global access to satellite and wireless signals offered by such behemoths as Hughes, General Motors, AT&T, Bertelsmann AG and other multinationals, with war chests brimming with billions of dollars, francs and deutsche marks. In company like that, Rogers looks especially puny.

Rogers no longer receives much support from the public. In a typically Canadian response to entrepreneurial success – or perhaps in response to a superior public relations campaign – they hold more sympathy for the faceless phone companies than for Rogers. Special-interest groups like the Friends of Canadian Broadcasting and the Consumers Association of Canada criticize Rogers for fighting only in a controlled environment of regulated industries like cable and telecommunications, where he can gouge the public with regulatory impunity. There's no doubt that regulations have enabled Rogers to compete directly with far bigger organizations without having his corporate block knocked off. The Bell Canada Act, for example, prevents the phone company from holding a broadcasting licence, nor can it control or influence the content or "the meaning or purpose of messages transmitted." This may ensure freedom of expression and prevent a stern-voiced Bell Canada operator from butting into a session of heavy-breathing phone sex. But it also prevents Bell Canada from competing with Rogers in the cable industry. Among other things, cable regulations actually require cable companies to produce community-based programming.

Since Canada's regulators change their collective mind as often as they change their socks, however, Rogers can't take much reassurance from the dubious protection afforded by Canadian rules. But he can use the regulatory maze to his best competitive advantage, always keeping in mind the admonition of his late mother, Velma: "Don't bang your head against a brick wall. Go over it, under it, around it. Anything but through it."

Rogers has gone over, under and around, but not through the brick wall of Canadian regulations, while maintaining a contradictory sense of competition. In his fight to enter the long-distance business, for example, he said, "Competition is good for the country, good for the customer, and what's wrong with that?"

But almost in the next breath, he has railed against the possibility of phone companies competing against him in the cable industry, at least until he can build Rogers Communications from a relative corporate weakling into the Charles Atlas of the industry. Even Rogers admits that this is an awkward position to defend. "It's like having four children," he said. "You tell each of them something a little different."

As technology evolves, however, the distinctions between one provider and another are becoming irrelevant, along with the regulations that protect them. In the process, Rogers's argument is rapidly losing its validity. As Canadians realize they can watch *I Love Lucy* on their PCs and make long-distance phone calls to Tasmania over the Internet, why should they patronize Rogers? In fact, most Canadians prefer the telephone company to Rogers. Telephone lines reach 98 out of every 100 Canadians, while cable reaches only 87 of every 100 Canadian homes served by the industry. For mysterious reasons known only to a public relations genius, phone companies also enjoy a reputation among the public out of all proportion to their performance relative to cable companies. By a margin of four to one among consumers and seven to one among businesses, Canadians have said they prefer the phone companies to the cable companies as providers of electronic information. Canadian broadcasters and publishers also oppose the Rogers monopoly in cable. In 1994, Izzy Asper, founder of Canada's third-largest TV network,

CanWest Global, urged Ottawa to allow satellite, phone and even microwave companies to distribute TV programming in competition with cable.

In the United States, 55 per cent of cable-TV households say they would likely abandon cable in favour of phone or satellite companies if prices were competitive. Regional Bell companies in the United States all plan to offer video programming, and almost 1 million households have purchased direct broadcast satellite dishes from companies like GM Hughes and Hubbard Broadcasting. Harold H. Greene, a U.S. federal district court judge who oversees the seven Bell regional operating companies, ruled in March 1995 that Bell Atlantic could compete directly with cable operators and TV broadcasters in transmitting video programming anywhere in the country. The ruling will soon be extended to the other Bells in the United States. With this in mind, Canada's regulators, who have long supported the cable industry's growth, now anticipate that phone companies in Canada will provide cable service by 1997.

That doesn't mean the public will subscribe, no matter who provides the service. Another survey, conducted in 1995, indicates that Canadians feel they have neither the time nor the desire for two hundred TV channels. Even if they didn't have to pay a penny more than their current cable bills, only one in five would sign up for such an expanded service.

Nevertheless, the phone companies have taken a page from Rogers's playbook to sneak over, under and around Canada's regulatory walls. "We're talking about competing in Rogers's territory," said Ray Cyr, BCE's president and CEO, following the company's annual meeting in 1989. And the phone companies have done just that and more. Since 1989, they've proceeded to put equipment in place and set up pilot programs to compete with cable services in Montreal,

Vancouver, Saskatoon, Yellowknife, London, Ontario, and elsewhere, while challenging regulators to shut them down.

As the years pass, Rogers continues to hold his ground, insisting that the cable industry cannot yet compete with the telephone companies. But while his arguments remain valid, regulators no longer take them at face value, and the public, if it cares at all, doesn't understand them. "If there is to be competition in cable television," Rogers said in 1994, "it must truly benefit consumers and be both fair and sustainable. The CRTC has a number of regulations, policies and practices for cable companies that determine which services they may or may not distribute and even how to package certain services. For competition to be fair, the telephone companies would have to live by the same rules as the cable companies, which would lead to virtually identical packages of services. But if both competitors are limited to providing identical packages of services, how does competition provide greater choice for consumers?" In response to this question, millions of Canadians scratched their heads, shifted in their chairs and channel-surfed from the O. J. Simpson trial to *Geraldo*.

But Rogers perseveres: "While we welcome competition from satellite and other providers, we believe that safeguards are needed if the phone companies build competing systems. Specifically, without safeguards, the [phone companies] could shift costs of the facilities to telephone subscribers, raising their rates and unfairly competing with cable."

With a monolithic presence in every area of the telecommunications industry, Rogers says, the phone companies can manipulate their accounting records to support any position they may choose. While cable companies are honest, upright, brave, courageous and bold, like Wyatt Earp, the phone companies "hoax" the regulators, he said in 1994, by exaggerating how much it costs to provide local phone service, where they have a monopoly. They "cook the books," then price those services where they do face competition below cost to drive their competitors out of business. "They don't intend to compete," warned Rogers. "They intend to annihilate."

Meanwhile, Rogers has urged the cable industry to consolidate even further to resist the phone companies' onslaught. "The most important priority," he said, "is the war with Bell. We must mobilize and close ranks, develop a national cable system."

Using a map of Ontario to justify his position to a reporter from the *Wall Street Journal*, Rogers pointed to the limited territories served by his cable company and others. But the telephone company is everywhere, he said. "You just have to know military history to know what will happen. We'll be encircled, and we'll be wiped out."

Added Phil Lind: "It would be like a fight between an 800-pound gorilla and a bunch of little ducks. You're not going to get much of a battle."

When reasoned arguments fall short, Rogers has resorted to colourful analogies to emphasize his determination to fight the enemy. After listening to a speech delivered by Bell Canada's former chairman, Jean de Grandpré, for example, he compared it to Hitler's *Mein Kampf*.

"The cable industry is not survivable in an open street brawl beginning tomorrow," Rogers told the CRTC in March 1995. Estimating that Rogers Cablesystems would lose 130,000 customers within two years to satellite services, he added that the phone companies "could teach Richard Nixon" a thing or two. Not one to lose the chance to offer her own mangled analogy, Jocelyn Côté-O'Hara, president of Stentor, replied that Rogers's comments bordered "on libellous, coming from a Watergate plumber."

Even in the days of Richard Nixon, Rogers had the phone companies in his sights, whether they knew it or not. When he entered the cellular industry in 1982, he was like Jack planting the beanstalk. But Rogers and his executives knew that their entry into the cellular industry marked the beginning of their head-to-head competition with the telecommunications giant. "From the moment we entered cellular," said George Fierheller, vice-chair of Rogers Communications and former president of Cantel, "that was the moment we took on Bell."

As the cellular industry took off, however, Rogers became a more threatening competitor. Soon he found himself competing not only with the phone companies but with his own organization, as well. The initial market for cellular phones had been conquered within five years of Cantel's formation in 1982. The salespeople who used a cellular phone from the front seat of their automobiles as they tooled around town peddling real estate and life insurance had quickly recognized the value of the technology. By the late 1980s, Rogers had to look farther afield to newer customers to keep his market and his company growing. These new customers wanted a cellular phone simply to shoot the breeze as they sat in traffic staring at the back of the driver's head in the car in front of them. But they didn't want to pay hundreds of dollars a month for the dubious delight of arguing about yesterday's dinner party with the spouse who was taking care of the screaming kids at home. To accommodate them more cheaply, Rogers began selling phones and luring subscribers directly. "It was exactly the same as if GM started selling cars from its Oshawa factory in competition with its own dealers," said Tom Ungar, a former Cantel dealer who joined the company in 1984 and sold his dealership back to Rogers Cantel in 1991. "My competition had always been the phone company. I still had them, but I found I had a new competitor as well – Rogers Cantel."

Even if he did step on a few toes in the process, Rogers Cantel had acquired more than 800,000 subscribers by 1995. With a barrage of advertising and with rates as low as $20 a month, new subscribers were signing up by the thousands. But many of them were choosing the phone companies over Rogers. Bell Mobility, the cellular service operated by Bell Canada, had almost 650,000 subscribers in Quebec and Ontario alone. Mobility Canada, representing the cellular affiliates of all Canadian telephone companies, had signed up almost 1.2 million subscribers by March 1, 1995. Once again, Rogers found himself in a battle with his arch rivals for survival.

Meanwhile, other technologies with names like personal cordless telephones, digital cordless telephony and enhanced specialized mobile radio were on the horizon. According to Rogers, "Such services will grow the overall wireless market, which will benefit all competitors." Rogers added that he "believes that regardless of the technology deployed, [Cantel] subscribers will be driven to purchase by the product quality, marketing and price, not by the technology deployed."

As Rogers laid out his cable networks and built his cellular system, he realized he had the foundation of his own phone company. By adding a few missing links, he could provide customers with an alternative to the telephone system that Canadians, for years, had loved to hate. And he knew exactly where to find the missing links. In 1989, Rogers bought 40 per cent of CNCP Telecommunications for $275 million and began to mount his attack on the phone companies' monopoly. Meanwhile, he began papering Ottawa from Bank Street to Rideau Hall with reams of documents explaining why the time had come for Canadian regulators to open the long-distance business to competition.

Not surprisingly, the phone companies resisted, fighting back in their customary fashion. Phone-company executives donned their best suits and said that if competition caused a drop in long-distance rates, then they would raise local phone rates, because the phone companies were using current long-distance revenues to subsidize local services.

To meet this argument, Rogers said the company he'd formed with CNCP, now called Unitel, would pay fees to the phone companies for access to their local systems. The fees, he said, would cover most of the lost revenues now applied by phone companies to local service from their long-distance charges. To compete aggressively, Rogers also said Unitel would forgo profits until 1997.

The phone companies pooh-poohed Rogers's pitch and said he was too weak to compete with them. In 1991, they said he and his partners could not afford to pay $13.5 billion over the next fifteen years to build a rival long-distance system. But Rogers claimed that the demand for long-distance services over that period would quintuple, with revenues reaching $26 billion by 2007. In such a market, he could easily scrape together a few billion from operating profits.

Meanwhile, consumer groups joined the phone companies to argue against competition in long-distance service. Consumers would ultimately lose if they had to pay higher rates for local service, they said. "If we have competition in the long-distance marketplace, the overall phone bill for 90 per cent of telephone subscribers will rise," said David McKendry of the Consumers Association of Canada. "The only people who are going to benefit," he added, "are the ones who make a lot of long-distance calls."

Canada's New Democrats also saw a gold-plated chance to display their concern for the little guy, whose welfare they aimed to promote by tarring and feathering the businesses that employed him. As Ian Waddell, the NDP communications critic in Ottawa, tooted from the

Rogers's father, Edward Samuel Sr., began building radios as a child and started CFRB – Canada's First Rogers Batteryless – when he was twenty-seven. "He was my hero and role model," Ted Rogers has said.

When asked in January 1953 by a U.S. immigration inspector if he was affiliated with a political party, twenty-year-old Rogers replied that he belonged to the *Progressive* Conservatives. By then, Senator Joe McCarthy had been conducting his witch hunts for three years, and progressives were associated with communists. The inspector detained Rogers and his friend, William Boultbee (right).
AP WIREPHOTO

The detention of Rogers and Boultbee, under the U.S. McCarran Act, prompted a demonstration in their support at the University of Toronto. There would be more demonstrations about Rogers to come, but this was the last held in his favour.
CANADA WIDE

Rogers and a colleague met Progressive Conservative leader John Diefenbaker when Rogers was president of the PC Student Federation at the University of Toronto. The federation eventually removed Rogers from the top job when he blew the organization's annual budget on a luncheon for the Chief. GLOBE AND MAIL

After a six-year courtship, Rogers married Loretta Robinson at St. Margaret's Church in London, England, on September 23, 1963. Loretta's parents, Sir Roland Robinson, MP, and Lady Robinson, are on the far right. At far left are Rogers's mother, Velma, and stepfather, John Graham. CANADA WIDE

After Rogers appeared before a federal inquiry into mass media in 1970, led by Senator Keith Davey, the Liberal government of Pierre Trudeau ordered an increase in Canadian programming on Canadian television and radio stations. CANAPRESS

Rogers has spent so much time at hearings before the Canadian Radio-television and Telecommunications Commission that he has referred to the organization as his second wife. In March 1995, CRTC chairman Keith Spicer (left) welcomes Ted Rogers to yet another hearing, this one on the information highway. CANAPRESS

Rogers would like to keep his empire in family hands. His son, Edward, worked for Comcast in Philadelphia when it bought Maclean Hunter's U.S. cable operations in 1994. His eldest daughter, Lisa, works for Rogers Cablesystems. CANAPRESS

A friend of the family who joined Rogers after graduating from university in 1969, Phil Lind played the role of court diplomat, dealing with Canadian community groups and regulators as Rogers expanded his cable system from one municipality to another. In the 1970s, Lind spearheaded Rogers's drive into the United States. Now vice-chair of Rogers Communications, Lind has worked with Rogers for almost thirty years. GLOBE AND MAIL

George Fierheller was a classmate and fraternity brother of Rogers at the University of Toronto who headed west to Vancouver to become president of Premier Communications. Rogers acquired the company in 1980, and Fierheller became chairman and CEO of Cantel. With Phil Lind, he is now vice-chair of Rogers Communications.
GLOBE AND MAIL

Graham Savage joined Rogers in 1986, replacing Robert Francis, who had helped Rogers for ten years in building Rogers Communications into a major national enterprise. When Francis died suddenly in January 1986 at the age of fifty-two, Rogers turned to Savage to keep the financial ship afloat.
GLOBE AND MAIL

Colin Watson never expected Rogers to succeed in his acquisition of Canadian Cablesystems Limited in 1978. But when Rogers took over the company, Watson stayed on board. He is regarded as one of the leading figures in the North American cable industry.
MACLEAN'S

Ted and Loretta Rogers were all smiles after Rogers announced that he had acquired
Maclean Hunter on March 8, 1994, and Maclean Hunter president Ron Osborne
seemed to share their enthusiasm. Canada Wide

But when the lights went up and Rogers executives gathered for a press briefing in
December after the CRTC approved the acquisition, there was no room on the team
for Osborne, who soon joined Rogers's arch rival, BCE. Globe and Mail

Ted Rogers in early 1994 at the age of sixty. Rogers's father died when Ted was only five. "It's very hard to compete with your father. It has always driven me."
MACLEAN'S

back of the regulatory bus: "Don't let these guys fool you. Under competition, local rates will go up, business will benefit and consumers will lose."

The debate continued. Rogers pointed out that in the United States long-distance rates had dropped by about 40 per cent in the five years after competition was allowed. The number of calls increased, and so did revenues to the carriers, while fibre optics helped to reduce costs. In short, everyone was happy. Rogers's opponents countered that local rates in the United States had risen by 50 per cent in the same period. Yes, but, Rogers continued, U.S. regulators attributed the increase not to long-distance competition alone, but to changes in regulatory policy that allowed U.S. phone companies to balance long-distance and local costs.

Besides, Rogers added, "competition stimulates development" and "Soviet-style communications monopolism is out of date. Even in Russia they're allowing some competition," he observed, as phone-company executives, consumer advocates and NDP critics goose-stepped around the Parliament Buildings looking for a pay phone.

Rogers's arguments prevailed. In June 1992, the CRTC allowed long-distance competition. But it was no gift to Rogers. In its decision, the CRTC opened the long-distance field to domestic and foreign-owned resellers of all shapes and sizes. Unitel became just one of many companies competing for a limited number of long-distance dollars. And since Rogers had promised that Unitel would set its rates 15 per cent lower than the phone companies, a price war was imminent.

The phone companies continued to moan and complain even as they prepared for the inevitable. They immediately appealed the CRTC's decision to Canada's federal court, while allying themselves with MCI Communications in the United States to gain access to critical technologies such as software for billing services and virtual private networks. They also waged a public campaign focused on Unitel's reputation for quality-control problems. And they continued to prey on public fears that local rates would rise in response to long-distance competition.

As a major shareholder in Unitel, Rogers found things were far from rosy. In arguing for long-distance competition, he had asked for a five-year period in which Unitel would make discounted payments to the phone companies. But after the federal court upheld long-distance competition, Unitel ended up paying more than 50 per cent of its revenues to the phone companies. Nevertheless, Unitel vowed again to forgo profits until 1997, a promise it has kept without even trying.

The phone companies found other ways to fight back. In 1994, they proposed a discount scheme that would give subscribers points for every call they made. They also expanded their local calling areas, eliminating the need for long-distance service within the expanded boundaries and eliminating competition in the process. Meanwhile, Unitel began competing against Rogers Network Services in providing high-speed private links to corporations such as banks and investment dealers. Once again, the whoopie cushions were waiting and the bananas were on the mat.

While Rogers fought the long-distance battle with one hand, he kept a grip on revenues at Rogers Cantel, which were threatened by falling long-distance rates. "Recent reductions in wireline long-distance rates," he observed in 1994, "will affect [Cantel's] long-distance margins. Wireless subscribers also have the ability to use alternate long-distance carriers by dialling access numbers."

By 1995, there were 150 companies competing with Stentor, the phone companies' national long-distance organization. Not one of them was profitable. Many companies, including Unitel, claimed that the phone companies were demanding such high access fees that no one could afford them. Regardless of the reasons, the phone companies still controlled 82 per cent of the long-distance market, leaving the other 150 competitors to bicker over $960 million in subscriptions while they paid half their revenues to the phone companies in access fees. To capture market share, companies slashed their prices. At one point, Unitel was offering Toronto-to-Montreal residential calling at night for 6.5¢ a minute, 4.5¢ of this to be given to the phone companies. "Anybody who thought you could enter this marketplace and it would be like shooting fish in a barrel was dreaming," said telecommunications consultant Eamon Hoey. "It's not for the faint of heart."

Even as long-distance competition went into high gear, it took place in a limited context. Unitel and its competitors must file all their rates with the CRTC and can't negotiate prices for special purposes, even for business services or private lines, as companies in the United States can. Nevertheless, even limited competition finally brought relief from rates that had been up to 40 per cent higher in Canada than the United States.

Naturally, the phone companies didn't take the competition lying down. By 1995, they were scrambling to raise local rates, and Bell Canada had announced plans to charge business users for each local call instead of a flat rate.

As the phone companies diddled with their local rates and services, Rogers mounted a drive to compete in that area as well. To anyone who would listen, including his shareholders, he called for competition in local telephone markets and the interconnection of local systems offering local switched voice services. "However," he added, "there are a number of issues, including number portability, interconnection arrangements and contribution, that need to be overcome before competition in local service can really happen.

"As a result of these regulatory issues, we believe there will have to be a period of transition before a sustainable model for competition in both cable and local telephone service can emerge. We hope that the CRTC will recognize the need for this transition period and for safeguards to ensure that there will be fair and sustainable competition in both the cable and local telephone businesses that can truly benefit consumers."

The gods of regulation were smiling on Rogers as he spoke. A few months later, in May 1995, the CRTC issued a sixty-one-page report on the telecommunications industry. Among other things, it called for competition in the cable industry, but only after the market for local telephone service was opened to competition. Once again, Rogers owed a small debt of gratitude to the regulators.

But Rogers knows he cannot rely on such regulatory reprieves indefinitely. As new technologies eliminate the distinction for consumers between TV, personal computers, telephones and the lines that connect them to the outside world, Canada's approach to regulation will become more absurd, archaic and counter-productive than it is already. Even the CRTC has sensed that its own days may be numbered.

Meanwhile, in television, Power DirecTV and Expressvu prepared to offer satellite broadcasting services, just waiting for the day Canadian regulators would give them the go-ahead. For Expressvu, the CRTC approval came on July 7, 1995. Both satellite companies were ready to begin full operation in Canada during the fall of 1995.

Initially, the CRTC prohibited Power DirecTV from operating in Canada because it would use a high-powered U.S. satellite to deliver its services. Under government pressure, the CRTC relented. Now both companies plan to start beaming their television services into Canada. Like its cable competitors, Expressvu will offer a basic service, a pay-TV service and a twenty-two-channel pay-per-view service. Satellite dishes will cost $900 to $1,000, although this price is expected to fall considerably as more people sign up. Rogers has said he expects the high initial cost of the dish will limit the number of satellite subscribers. But for the 2.4 million homes in Canada not served by cable, satellite service will come as a welcome relief from reruns of Peter Mansbridge blathering on about VE Day in Britain. Rogers also knows that in the United States, DirecTV sold 600,000 dishes in eight months, and 100,000 of those purchasers dropped their cable service in the process. A 1992 study conducted for the Canadian Association of Broadcasters predicted with astonishing precision that "4.6 per cent of Canadian basic cable subscribers [would] disconnect by the year 2000" because of satellite competition. Two years later, even Rogers admitted that about 200,000 of his 2.4 million cable subscribers would buy dishes in the first couple of years after they became generally available.

To compete with satellite and provide subscribers with a similar array of selections, Rogers Cablesystems has expanded its pay-per-view service to twenty channels from four. Rogers also hopes to introduce video on demand in 1996, if he can obtain converters, headend equipment and other geek machinery required to get the service up and running. Initially, he has pledged to install 200,000 of these boxes in Vancouver, at a monthly rental fee of $7.95.

Rogers is less concerned with the number of channels carried by satellite competitors than with the regulations governing their content. "If Power DirecTV is allowed to offer services in Canada and deliver the full array of services offered by DirecTV in the U.S., with no restrictions on content or packaging, this will pose a serious competitive threat to cable television," he said in 1994.

Further competition for cable may come from multipoint distribution systems (MDS), which use microwave transmission instead of cable to deliver packages of channels and services to subscribers. Because its network of antennae and receivers have to be within open view of each other, MDS can operate only in populated areas. Regulators have refused so far to sanction MDS, but that could change with a nod and a wink from Keith Spicer.

Rogers has even more tricks up his sleeve. Installing technologies such as interactive cable services, Rogers will soon provide subscribers with games such as SegaVideo and high-speed data services, while enabling the power company to read a subscriber's hydro meter over the cable line. The services will be delivered over television, PC and automatic meter devices.

All this depends on keeping the phone companies out of cable for as long as possible. True to his legal training, Rogers continues to split hairs to find new reasons to defend his position. Here's one: There's vast room for expansion in the long-distance market, he has said, but everyone who wants cable TV already has it. The additional costs of competition would not bring additional revenues to the industry, and everyone would do poorly.

With this in mind, Rogers has urged the CRTC to deny the phone companies' $8-billion proposal to install fibre-optic lines in Canadian homes. "If there's going to be one fibre-optic cable into the homes in the Rogers areas, it's going to be Rogers's," he has said. "It's not going to be someone else's." Copper wire provides all the capacity that phone companies need to handle phone calls, he says. And the cost of installation of fibre-optic line by the phone companies would likely be borne by consumers. So far, Rogers is a prophet without a country. The consumers' associations that once rallied around the phone companies to protect consumers from higher costs while opposing

Rogers's entry into the long-distance market now remain stunningly silent. Meanwhile, the Friends of Canadian Broadcasting, who would regulate your VCR in the name of Canadian culture, are clamouring for competition. The Friends admit, as Rogers pointed out long ago, that "one of the major benefits of competition in [telecommunications] is the incentive to invest in further technological research and development to provide customers and end-users with innovative services and products."

With a realistic sense of paranoia, Rogers expects phone companies to initiate a deal with the cable industry as they lay their fibre-optic networks, allowing some of Canada's two thousand cable systems access to phone companies' lines in return for a fee. "If I was running the phone company," Rogers says, "I would have wined and dined and seduced the cable industry and offered to rent video fibre-optic cable at relatively small amounts for the first five years, then gradually increased the rental so that I wiped them out and took over their operating profits.

While Rogers would prefer consumers to use his cable system to handle local phone calls, cable service and long distance, others see the future differently. Peter Huber, at the Manhattan Institute, predicted in May 1995 that the information age would be consummated through dual conduits into the home – a telephone line and a cable connection. "We are on a two-wire track," he wrote in *Upside* magazine. "Phone companies and cable companies will be allies. It's not the cheapest way to go, but the price you pay for a monopoly is exorbitant in the long term. You lose innovation, and you lose all the discipline of competition. We are going to a head-to-head battle between cable and telephony, both providing video, both providing voice, both using wireless for the last hundred yards."

Mindful of the possibility of such dirty tricks as subsidizing the competitive activities of their local phone services, pricing below cost and denying access to their poles and ducts to cable operators and long-distance competitors, Rogers wants the telephone industry to set up separate companies to run their local phone and cable businesses, relinquish control of their long-distance companies and dismantle the Stentor consortium. "If the CRTC requires the telephone companies to offer competitive services through a separate company, then the opportunity to misallocate costs is diminished. We have seen how well this can work in cellular telephone competition, where Bell and B.C. Tel have set up separate companies to provide cellular service. As well, in the United States, where the local telephone companies are separate from the long distance companies, competition in long distance has been a great success."

Meanwhile, Rogers continues to press for consolidation in the cable industry. "The fundamental flaw for Canadian cable is that it's fractured," he said. "When you talk about a satellite that serves all of North America, for the cable companies to resist is very awkward. Without consolidation, acquisitions, mergers or swaps of territories, how can you possibly negotiate fibre optics and other things on an efficient basis?

"It makes no sense for Hamilton to have four or five different cable companies or for Toronto to have three or four. I don't mean that one company should own all the cable systems across Canada. What I mean is you'd have all of Edmonton or all of Calgary owned by one company. If I'm going down one side of the street and another company's going down the other side, it's virtually impossible to come up with a low-cost efficient method unless you have consolidation.

"If you're going to deal with competitive threats from satellite or from the telephone company, you have to have an industry that's

In return for their loyalty, Rogers has rewarded his employees handsomely, not just financially but emotionally as well. He attracts competitors who like to win, and so far they haven't been disappointed. Besides, winning is always fun, and his closest allies have stuck by Rogers for decades.

At critical times, Rogers has also formed strategic alliances with partners outside the organization, initially with his neighbourhood cohorts, the Bassetts and Eatons, and later with allies farther afield. In fact, the larger his organization grows, the more he needs partners to help him, particularly in competing against telephone companies in Canada and other countries. Although Rogers Communications now ranks among the top sixty-five companies in Canada, Rogers continues to motivate his troops by emphasizing their corporate role as the underdog of the telecommunications industry. "It's always an uphill battle," says vice-chairman Phil Lind, a Rogers employee for more than thirty years, "always a struggle."

Until he acquired Canadian Cablesystems Limited and Premier Communications between 1978 and 1980, Rogers relied most heavily for inspiration, guidance and execution on one corporate executive, Phil Lind, and a handful of close relatives and associates, including his wife, his mother and his stepfather. A friend of the family who had joined Rogers right after graduating from university, Lind played the role of court diplomat, dealing with Canadian community groups and regulators as Rogers expanded his cable system from one municipality to another. In the 1970s, Lind spearheaded Rogers's drive into the United States, cajoling and persuading his way from Portland, Oregon, to Minneapolis to Los Angeles, acquiring cable licences and concessions at each stop.

With the acquisition in 1978 of CCL, Rogers added Colin Watson to his team. Reputedly the most knowledgeable cable industry executive in Canada, Watson had fought to keep Rogers from acquiring CCL. But when the time came, he readily joined the Rogers team. An innovative competitor who chugs down the streets around his Rosedale home on a Harley Davidson when he isn't jogging or playing tennis, Watson guided Rogers Cablesystems through the 1980s, turning it into a technological powerhouse and acquiring 2.3 million subscribers along the way.

George Fierheller had known Rogers at university and joined the company in 1980 when Rogers acquired Premier Communications, where Fierheller had been president. With encouragement from Fierheller, Rogers entered the cellular industry two years later and appointed Fierheller as president of Cantel. Under his guidance, Cantel had blossomed into a $493-million-a-year business by the time he retired from the position of chairman and CEO in 1993.

By that time, Rogers himself was pushing sixty, had hired all the capable friends he could find for executive positions in his company and wasn't prepared to acquire another company in the cellular industry just to capture a capable executive. To replace Fierheller, he found David Gergacz in Massachusetts. An engineer and economist, Gergacz had been president of Boston Technology Inc. in Wakefield, and had experience as head of network systems for Sprint Communications, a leading long-distance carrier. He began his career at Bell Laboratories in Murray Hill, New Jersey.

Just fifteen years earlier, Rogers might never have attracted a leader of the calibre of Gergacz. "Before 1978," Rogers says, "my problem was that I had all these great visions, but we didn't have the strength to do them right. We now have a visionary team and experienced professional managers with a strong technical background."

Many of them came to Rogers like prizes in a box of Cracker Jacks from companies that he acquired. If they were talented, hard-working and ambitious, Rogers kept them on his team. Even after protracted battles for control, Rogers "is not one to settle old scores," says Lind. Although the top man usually departs, as Ron Osborne did after the battle for Maclean Hunter, joining BCE, and Tony Griffiths did after the fight for CCL, Rogers does not force him out. In fact, he usually invites him to stay. Meanwhile, Rogers retains much of the team that makes the takeover target so successful to begin with.

Even with more than 13,000 employees, RCI is far from a lumbering bureaucracy. Few people in the organization occupy a desk until they collect a gold watch and a pension at the end of their term. In fact, the organization could probably use a few more lieutenants, as Rogers discovered suddenly when CFO Robert Francis died of a heart attack in January 1986 at the age of fifty-two. Francis had worked with Rogers for more than ten years, guiding the company through economic ups and downs, including a recession in 1982 and 1983. In Francis's sudden absence, Rogers assumed the duties of CFO and dived into negotiations to sell his company's U.S. holdings to reduce its $900-million debt. "If Ted ever had a partner, it was Francis," said Jim Sward, then president of Rogers Broadcasting. "Francis was the fellow he worked with, dreamed his dreams with and chased the banks with."

While struggling with the company's debt and negotiating the sale of his U.S. assets, Rogers was also reorganizing his $1-billion company for the future. In April, the company changed its name to Rogers Communications Inc. from Rogers Cablesystems Inc. In May, it purchased 46 per cent of Cantel from Rogers's private holding company for $35 million. In June, it bought 80 per cent of MTV, a multilingual broadcasting company in Toronto. In the meantime, Rogers was buying, selling and trading his U.S. cable holdings.

As the company expanded, Rogers added more close associates to his board, such as Gordon Gray, former chairman of Royal Lepage and now chairman of Rio Algom Limited; John Tory, deputy chairman of Thomson Corp.; Chris Wansborough, president at the time of National Trust and now chairman of OMERS Realty Corporation; and Bob Smith, who was president of Talcorp, a private investment management firm.

"We don't have much turnover here in management," Rogers said. "We do have strenuous debates, sometimes at the board level. I don't win all the votes, even though I control the company. Graham Savage has won votes. No one has ever been reprimanded or fired for causing me to lose a vote.

"People who suck up are of no value," he continued. "I pick people who stand up for what they believe. Close arguments, fighting back and forth, usually produces the best decisions. I can assure you I don't always win."

Although he may not win all the battles, there's no doubt about the victor in any internal war. "We terminate people if they don't produce what they say they will," Rogers said. "We're very tough on that."

Added Phil Lind: "If you are not seen to be relevant to the effort, then maybe you're not going to be in on the effort."

Although he may be a demanding boss, Rogers needs his sidekicks almost as much he they need him. "He's a superb visionary," says Philippe de Gaspé Beaubien, chairman of Telemedia and a strategic partner of Rogers in the early days of Cantel. "He's a great risk-taker and an entrepreneur. But I think Ted would be the first to say that he is not a manager; he is someone who hires good managers and works well with them."

With the help of his seasoned cable warriors, Rogers marched onward. And the bigger the company grew, the richer they all became. Now, after thirty years of playing Let's Make a Deal with

Rogers, Phil Lind holds more than 335,000 Class B non-voting shares and 190,000 Class A voting shares of Rogers Communications. Graham Savage, who replaced Francis as chief financial officer and senior vice-president of finance, holds more than 216,000 Class B shares. Watson holds 254,000 Class B and 147,000 Class A voting shares. More than a dozen managers have made millions from their holdings in the company. (Rogers himself holds more than 50 million voting shares and 9 million non-voting shares of the company.) Money can't buy loyalty, but it comes in handy when the boss is on the phone at 8:30 on a Saturday morning just as you're packing the kids in the car to drive to the cottage.

While his executives have brought their operating skills to the company, his directors bring their power and political connections. Even though Rogers is a lifelong Conservative, counting among his friends Brian Mulroney, with whom he first worked during the leadership campaign of John Diefenbaker in the 1950s, a number of Liberals sit on Rogers's boards. They include David Peterson, former Liberal premier of Ontario, and Francis Fox, former minister of communications in Pierre Trudeau's government, who is a director of Cantel. Fox was communications minister in 1983, when the federal government turned down CNCP's application to compete in the long-distance industry. Long before he joined the board, he was familiar with Cantel, since he had approved Rogers's application for a licence to operate the company as a national mobile telephone service. In the process, he had defended his decision against several Liberals from Toronto, who blew their head gaskets at the idea of a Tory like Rogers receiving any favours from a Liberal government. In fact, Fox saved Cantel from annihilation at the hands of the phone companies by refusing to give them a head start in the market, even though they were prepared to get their own mobile phone service off the ground several months ahead of Rogers. "The telephone companies would have killed us if they'd all started at once everywhere," said George Fierheller.

Other connections have come in handy for Rogers as well. Pierre Juneau, former head of the CBC and former chairman of the CRTC, oversaw the affairs of Maclean Hunter in trust until Rogers received approval of his acquisition from Ottawa.

Sometimes the connections of his directors and advisers work against Rogers. In 1994, for example, Peter Eby, vice-chairman of Burns Fry, and Gar Emerson helped Rogers put together his offer for Maclean Hunter. Five years earlier, Eby and Emerson had helped Maclean Hunter's board draft its poison pill, a mechanism designed to make a takeover less attractive. When Rogers made his offer, Maclean Hunter's directors threatened to take legal action "to ensure that [Rogers] does not benefit from Maclean Hunter's confidential communications with and advice received from" the two executives.

In the midst of Rogers's allies sit the relatives, including Loretta, his wife. Rogers's stepfather, John Graham, was a valued director and chairman of the board for years. When Graham retired in 1993, Gar Emerson stepped in. Emerson was president of Rothschild Canada Inc., an investment bank set up by Rothschild & Sons of London, and had been a mergers and acquisitions lawyer at Davies Ward & Beck, serving as chief takeover adviser to the Reichmann family. He is also the husband of Rogers's first cousin.

Beyond the company, Rogers formed his first significant alliance with the Bassett and Eaton families in the 1960s, when Rogers first invested in CFTO while his friends and neighbours chipped in to buy a share of Rogers's fledgling cable company. The friendship and the businesses have lasted for thirty-five years. Rogers recently invested $13 million through his personal holding company to buy an 8 per cent stake in Baton Broadcasting, run by the Bassetts. Meanwhile, Eaton's department store is a potential customer for Rogers's home-shopping network.

On occasion, Rogers has also formed partnerships or relied on the support of such corporate luminaries as the Bronfmans, the Belzbergs, Mike Milken, Philippe de Gaspé Beaubien, Craig McCaw and Canadian Pacific's Bill Stinson. Each relationship has had a specific purpose, and none has lasted longer than necessary. "We have always considered broad-based strategic alliances difficult," says Lind. "Common benefits, strategies and goals are just too hard to pin down. Rather we have seen more potential in directed, project-oriented partnerships."

As described in Chapter 2, in the mid-1970s Rogers initiated a relationship with Peter and Edward Bronfman, through Edper Investments. Together, they held a majority of shares in CCL, and they struck a buy-sell agreement that would allow one party to buy out the other and assume control of the company. The relationship lasted only until Rogers exercised the buy-out clause and took control of the company.

In 1982, Rogers formed a partnership with Marc Belzberg, president of First City Financial Corp., and Philippe de Gaspé Beaubien of Telemedia, publisher of TV *Guide* and other periodicals. Rogers joined the partners with a personal $2-million investment after his board refused to sanction the entry of Rogers Cablesystems into the cellular industry. "We have a Catholic from Quebec, a Protestant from Ontario and a Jewish entrepreneur from Vancouver," Rogers quipped at the time.

When the partners and their advisers met in Ottawa to brief the Department of Communications on their application, the session began like a session in genealogy. As George Fierheller recalled, "Ted began his remarks by talking about his dad and radio; Belzberg told them about his grandfather and their piano business; and then de Gaspé Beaubien boasted about how his family arrived in Quebec in the early 1600s." The anecdotes prompted one DOC official to retort, "I've only been here for eight months."

Rogers subsequently put another $5 million into the partnership. Then he began buying out his partners. By 1989, he had acquired Cantel for $670 million, or $34 per potential customer. By the time he took control, Cantel covered a potential customer base of 18 million people, although it had only 150,000 subscribers. On revenues of $144 million, Cantel was generating operating profits of $25 million. Rogers had acquired another cash machine.

In 1994, Rogers reacquainted himself with Telemedia when he discussed selling Maclean Hunter's radio stations to the Montreal-based broadcasting and publishing company. He also talked to Shaw Communications about swapping cable holdings after the Maclean Hunter deal and to Quebecor to negotiate a deal for Maclean Hunter's state-of-the-art printing facilities. By the time the serious discussions started over Maclean Hunter, Rogers had his ducks in place. Not only did he know where he could go to unload some of Maclean Hunter's assets, but he effectively eliminated these companies from bidding against him.

Rogers seldom bothers to climb into a telecommunications vehicle with a partner unless he can drive. But not all his partners are content just to come along for the ride. Canadian Pacific, for example, was hardly a corporate wallflower in 1988, when Rogers first approached to acquire Canadian National's 40 per cent share in money-losing CNCP Telecommunications; and CP's chairman and CEO, Bill Stinson, wasn't about to swoon over some guy who'd never gotten his hands dirty in a switching yard. Quiet, introverted, tough and a model bureaucrat, Stinson was a fourth-generation railwayman and the son of a CPR claims investigator. At fifty-eight, he had worked for CP for more than thirty-five years when Rogers showed up. Although he worked in his father's business and was only a year older than Stinson, Rogers was a loquacious risk-taker with a larger appetite than CP for debt. Already he had ruffled some feathers by saying he might sell part of his share in the CNCP partnership to pay down Rogers Communications's enormous debt.

CP had lost more than $900 million in 1991 on revenue of $10.6 billion. It lost another $19 million in the first six months of 1992, and Stinson couldn't have relished the idea of risking even more money by helping his new-found partner take on the phone companies. But in public, at least, Stinson referred to the "synergy" that Rogers brought to Unitel.

Under this arrangement, Rogers appointed four of the ten directors on CNCP's board, and Rogers himself became chairman. More important, Rogers Communications representatives would occupy half the seats on the company's executive, and would audit committees and hold veto powers over major investments and other decisions, such as whether to bring in new shareholders.

Although some people speculated that Stinson wanted out of the partnership, the alliance breathed new life into the company that became known as Unitel. Even if he didn't enjoy complete control, Rogers gave the company renewed direction and purpose. "In the past, the management was terrible and seemed unable to decide on investments," said one analyst at the time. "CN was always hard-pressed for cash and CP was always wanting to invest in the business."

Not only did CP live up to its obligations as partner, it went even further. In 1990, as interest rates soared and Rogers found himself once again clinging to a leaky life-raft of floating-rate debt, Stinson helped Rogers with a temporary but invaluable injection of cash, keeping Rogers from drowning while he replaced $3 billion in bank debt by selling bonds and shares.

As with his other strategic partnerships, this one included a mechanism that would allow Rogers to acquire control of Unitel from his partner under the right conditions. By triggering a shotgun clause that would come into effect in September 1994, Rogers would have the right to acquire CP's share, at a price named by Rogers. Alternatively, Rogers could sell his holding to CP at the price offered or 135 per cent of the shares' fair market value, whichever was lower.

By 1992, Rogers had $2 billion in untapped bank credit lines. Analysts estimated that CP's 60 per cent share of the company was worth at least $500 million, even before considering the CRTC's approval of long-distance competition. But Rogers and his executives denied that they planned to exercise the shotgun clause. "They're good partners," Graham Savage insisted. "We're enjoying the relationship with them."

When AT&T acquired a 22.5 per cent interest in Unitel in 1994, Rogers began an even more promising relationship. Not only did AT&T provide Unitel with its latest CEO, it also gave the company the technological depth and international scope it needed to compete with Canadian phone companies. As Lawrence Surtees points out in *Wire Wars: The Canadian Fight for Competition in Telecommunications*, AT&T's strategy is typified by "its mission to carry any information, anywhere, anytime. Surtees believes that "AT&T is one of the few companies in the world with the resources to deliver on that promise." AT&T has also made the ability to deliver video the linchpin of its corporate strategy. "Video will be to communications in the nineties what facsimile was in the eighties," said John Mayo, president of AT&T Bell Labs.

From technology to consumer finance, AT&T has the muscle to

manufacture, market, deliver and finance all the necessary components to run the electronic highway right into your bedroom. Among
other gizmos and doo-dads of the electronic age, Bell Labs is developing the technology for delivering, storing and manipulating digitized
video signals in the home. "The impact on AT&T's global partners,
including Unitel, will be profound," Surtees continues.

For AT&T, Unitel provides another foothold outside the United
States, where it can expand its international business. Involved in a $1-
billion joint venture to offer phone service in Mexico, AT&T would
like to build a stronger relationship with Rogers to secure its dominance in the North American telecommunications market.

Occasionally, Rogers rubs his partners the wrong way. Recently,
without asking AT&T for permission to pluck a star from its corporate
firmament, he hired Stan Kabala to become president and CEO of
Unitel. Kabala had spent twenty-nine years at AT&T and was a valued
executive. In an apparent huff over the slight, an AT&T executive left
Unitel's board of directors. Kabala's subsequent performance at
Unitel, however, more than justified the consternation caused by his
hiring.

Under less scrutiny than his relationships with AT&T and CP,
Rogers's arrangement with Astral Communications (whose board
members include Phil Lind and Francis Fox) and TSN Enterprises has
yet to live up to its promise. The three companies started Viewers
Choice, a pay-per-view channel serving about 500,000 homes in east-
ern Canada that are equipped with a descrambler. While the technology has yet to prove itself, pay-per-view services in the United States
are building a market by running bare-knuckle brawls between karate
experts, kick boxers and street brawlers named Bubba in rings surrounded by chain-link fences.

Such alliances provide Rogers with the strength his company

needs to survive as a distinct Canadian entity in the communications industry. "You [need] a lot of relationships internationally," he said. "We have relations with Bill Gates at Microsoft and with AT&T and a number of others. You have to have size to do that."

With his inimitable golly-gee enthusiasm for entrepreneurs and technology, Rogers describes Gates as if he just discovered him in his kitchen tinkering with the dishwasher. "Gates to me is a visionary and a genius. He definitely has some very important ideas that we believe in and are following, and I think there will be an opportunity for us to work together."

One of their shared ideas is "far off the wall," according to Rogers. "[Gates] has bought the rights to a lot of the masters of painting. He is going to be able to, by a switch, watch reproductions in absolutely splendiferous colour, almost better than the originals, on special types of screens in any room in his new house. Think about that for cable," Rogers enthused like the Don Cherry of technology. "If we could get those rights and somehow develop a method of distributing it so that millions of Canadians would have an opportunity to see pictures almost better than the originals, when in their lives could our people see that?"

Before acquiring Maclean Hunter, Rogers had arranged with Microsoft to use the company's Tiger software for two-way set-top boxes, which would allow viewers to talk back to their televisions even more than they do now. The set-top box will store video and audio signals delivered to the home. Rogers also expects his long-distance and wireless telephone networks to enter the living rooms of his Canadian subscribers in 1996.

In five meetings to date, Rogers and Gates have discussed the rollicking marriage of computer, cable, TV and telephone technologies. "I've talked to Ted Rogers about what we're doing," Gates said in 1993 during a visit to Toronto. "He's very interested in investing in these networks."

Rogers and Gates also met to discuss a pilot project in Vancouver that would link hospitals and offices in the city. "Our talks are confidential," Rogers said at the time, "so I don't want to blow his cover. But it will be new and done in Canada first."

Gates, who also has a close relationship with Tele-Communications Inc. of Denver, Colorado, the largest cable company in the United States, said, "Rogers is an entrepreneurial company," and added that he'd like to work with the phone companies, too. Through Unitel, Rogers could help to make Gates's wish come true.

"I don't think Gates is meeting Ted as Rogers the broadcaster," said Phil Lind. "The issue here is the electronic highway and whose personal navigation system is going to be employed."

"Gates is like a prophet at the top of the Himalayas," Colin Watson added. "You go and talk to him and take away what you want from it."

With less fanfare, Rogers has developed relationships with other major companies to develop or gain access to new technologies. "We try to meet regularly with other people in different countries in communications, in cable, in the telephone business," he said.

In 1992, Rogers began working with IBM Canada, testing the highest-capacity fibre-optic connection in Canada at the time. The line can move data at the rate of one gigabit – one million binary digits – per second, the equivalent of moving over 62,000 pages of text every second.

"We are also teaming up with Intel and General Instruments," Rogers said, "to define and develop capabilities for delivering high-speed data to personal computers through cable modems, at speeds up to a thousand times faster than a conventional telephone modem."

Meanwhile, Rogers meets occasionally with Craig McCaw, a

cable pioneer and founder of McCaw Cellular Communications Inc. of Seattle. "McCaw and Rogers have had a long relationship," Rogers said. "He built his cable business while I was building mine. We are both in broadcasting, and our fathers were in broadcasting. There are many similarities." (In the United States, McCaw is venerated as a visionary, risk-taking entrepreneur who sold his company two years ago to AT&T for more than a billion dollars. In Canada, Rogers is regarded as a slavering vandal who would trash Canada's cultural heritage of stuffy panel shows, programs on the maintenance of begonias and televised sing-alongs with one sweep of his electronic shillelagh. Even Canada's financial community, who should know better, regards Rogers with apprehension simply because of or despite his success.)

Through Cantel, Rogers holds a 10 per cent stake in Claircom Communications Inc., a subsidiary of McCaw Cellular that provides a fully digital air-to-ground service that can send and receive voice and data messages.

Rogers is also talking to Apple Computer. In fact, Lind observed, "everybody is talking to everybody, because we're all trying to figure out where it's all going."

As he manoeuvres among the quick-moving monkeys and the lumbering elephants of the telecommunications jungle, Rogers maintains that the next wave of progress in the industry may have nothing to do with any of them. "The key competition will come from young entrepreneurs and dreamers," he said, "people who will work nights and weekends and build their businesses and create a nation." Perhaps with this in mind, Rogers has formed some small but potentially rewarding alliances with a group of entrepreneurial firms. Through Unitel, for example, he holds a share of Focus Technologies Networks of Mississauga, Ontario. FTN provides access across the country to the Internet, focusing primarily on corporate customers like National Grocers. With forty employees, FTN has developed software to provide simple access to the global network. "Our vision is to become the Unitel or Bell Canada of the Internet world," said Kevin Brook, FTN's director of sales and marketing.

Though the future may belong to entrepreneurial insomniacs with a flair for technology, the present belongs to much bigger companies, like Bell Canada. In fact, the most sensible but unlikely alliance that Rogers could make would involve the Canadian phone companies, his arch competitors. Such a relationship would give both parties access to a sophisticated and comprehensive network of fibre-optic lines, switches and transmission facilities, enabling them to compete with much larger multinational rivals.

Other phone companies throughout the world have already joined forces with major computer makers, cable companies and chip manufacturers. British Telecom, for example, is working on interactive television with Apple and Oracle. Thomson Multimedia in France has developed interactive services with Sun Microsystems. Deutsche Telekom and France Telecom tried to form an alliance with Sprint in the United States, the third-largest long-distance provider in the country. Walt Disney is forming alliances with three U.S. phone companies. Teleport Communications, which provides private phone service to companies like Merrill Lynch, Dow Jones and the New York Stock Exchange, is now owned by four major U.S. cable firms, three of whom joined forces last year with Sprint to offer cable and telephone services – local and long distance, wireless and wired. Teleport is expected to generate US$1.5 billion in revenues by 2004.

In Canada, Bell has alliances throughout the world, including a $25-million deal with Brazilian Abril Television to expand cable television services in São Paulo. Certainly, it recognizes the advantages that could be gained from an alliance with Rogers.

A number of obstacles stand in their way, not the least of which is their bitter twenty-year rivalry. Regulations also prevent an acceptable partnership, but their corporate cultures pose less of a problem. After all, if Rogers can successfully align himself with a lumbering behemoth like CP, then a relationship with Bell Canada would seem idyllic by comparison.

Whether the future belongs to the big fish or the small fry, all of them will, like Rogers, need help from one critical ally – the bank. As Rogers knows from experience, no one in Canada can build a company without eventually walking into a bank with hat in hand to pledge the house and the first-born child in return for a fully secured loan.

Rogers's relationships with Canada's banks began when he acquired his second radio station, using money borrowed from the Bank of Montreal. In 1981, he borrowed $152 million from Toronto Dominion Bank to buy UA-Columbia Cablevision, securing the loan with his radio stations and his stock in Rogers Cablesystems. He now sits on TD's board. Robert Korthals, former TD president, who sits on Rogers's board, says his indefatigable former customer goes through account managers like an athlete wearing out his running shoes. In fourteen years, Rogers has left twenty account managers panting by the wayside.

For Rogers, Canadian banks aren't just his partners – they're his life-savers. Without them his company would never have survived.

CHAPTER 7

Financing

Look, Ma, no hands

Great moments in science: Einstein discovers that time
is actually money.

GARY LARSEN

In late 1994, Ted Rogers rumbled up Bay Street in a dump truck full of
cash, stopped at College Street and shovelled out $3.1 billion to buy
Maclean Hunter. Of the total amount, $2 billion came in the form of a
loan from a group of Canadian chartered banks; $740 million was cash
raised in previous borrowings; and $270 million came from an issue of
convertible preferred shares. Rogers also used a $148 million revolv-
ing-credit facility to pay the interest on the loan and other expenses.

Once the deal went through, Rogers sold several Maclean Hunter
assets to pay for it. These included the company's European publishing operations, for $120 million; a number of radio stations, including
CFNY-FM in Toronto, for $27 million; and Maclean Hunter's U.S.
cable operations, for $1.67 billion. He used the proceeds to repay the
bridge loan and revolving-credit facilities.

Rogers also swapped cable systems with Shaw Communications Inc., raising another $201 million. He raised another $308 million by selling Maclean Hunter's maritime radio operations, for $18 million, and its printing operations, for $124 million; the majority of Maclean Hunter's U.S. publishing operations, for $111 million; and its Canadian business forms unit, for $53 million. The sales continued. Davis+Henderson, which makes cheques, and CFCN, a CTV television network affiliate in Calgary, brought in between $130 million and $150 million.

Meanwhile, Rogers folded Maclean Hunter's Canadian cable operations into his own company and combined the Toronto Sun, Maclean Hunter's periodicals division and four radio stations in Kitchener and Ottawa into a new subsidiary called Rogers Multi-Media Inc.

On a consolidated basis, Rogers Communications Inc. achieved revenue in 1994 of $2.25 billion, up 68 per cent from the previous year. Even excluding Maclean Hunter's contribution, the company's revenues rose 16.5 per cent. More important, operating income before depreciation and amortization increased to $711.1 million, up almost 60 per cent, a 30 per cent rise beyond Maclean Hunter's contribution.

As usual, though, the company lost money. Its net loss in 1994 amounted to $126.1 million. That was better than the previous year's loss of $182.4 million. But if Colonel Maclean were to see such a figure, he'd do the Watusi from the Great Beyond.

"RCI is in a better financial position today than before the merger," Rogers said in the company's 1994 annual report, "benefiting as it did from the substantial earnings power of the Maclean Hunter businesses, its underleveraged balance sheet and asset sales of non-core businesses at excellent sale prices.

"At the close of 1994, the balance sheet's debt-to-cash-flow ratio was actually better than at the end of 1993. With the retained operations of Maclean Hunter, RCI will be able to reach free cash flow and net income targets sooner than would have been possible without its contribution."

Graham Savage, Rogers's senior vice-president of finance and chief financial officer, explained further that "in the final analysis, we paid approximately $3.5 billion for the debt and equity of Maclean Hunter and have sold or will still sell assets approximating $2.2 billion, for a net investment of approximately $1.3 billion. Based upon 1994 results, we estimate that we paid about eight times operating cash flow for the operations we will be retaining and approximately $1,200 for each cable subscriber." When the smoke cleared and the dust settled, the newly combined companies laid off a total of 135 employees, contrary to the yelpings of Rogers critics everywhere. Rogers Communications now employed 13,300 employees, about one-ninth the number employed by BCE, but still far too many to invite to a backyard barbecue, even in Forest Hill.

Even for an Upper Canadian blueblood, Rogers had come a long way on an initial investment of $85,000. Other companies, including some that he'd acquired over the previous thirty-five years, could have done what he did, and perhaps done it more easily and with fewer sleepless nights, churning of the innards and gnashing of the teeth. But none of them had the momentum or, to be fair, the approval of their shareholders, to take such monumental risks as Rogers. The pension funds and retired pressmen who owned shares in Maclean Hunter, for example, would not have appreciated the company's managers jeopardizing their investment on a hunch about telephones that resembled Dick Tracy two-way wrist radios. These companies were an accountant's dream. Everything balanced, nothing was out of place, risk was minimal and all the lines rhymed. But they were an entrepreneur's nightmare. While these companies were content to write a technological sonnet, Rogers aimed to compose a sprawling epic.

Friends like the Eatons and the Bassetts, with more financial ink that your average next-door neighbour, have helped him write it. So has his wife, Loretta, an heir to the Woolworth fortune. But even her $16-million inheritance doesn't go far when it takes $500 million to set up a cable network or $3 billion to buy a company like Maclean Hunter.

Rogers has needed deeper pockets than his wife, family and friends could provide. Through bank loans, junk bonds, equity issues and preferred shares, Rogers keeps his organization afloat while piling on new debt-laden assets that initially threaten to sink the ship. Meanwhile, like a typical entrepreneur, he continues to put his personal holdings at risk, just as he did when he bought CHFI in 1960, when he started Cantel in 1982 and when he started Unitel in 1992.

To take on Maclean Hunter in 1994, Rogers and his wife risked their entire $1.2-billion stake in Rogers Communications. In the process, Rogers acquired a venerable company with a debt rating of A plus, using shares of his own company, whose debt rating has never been higher than B minus. Maclean Hunter was no longer a suitable investment for little old ladies who have retired from teaching in Wawa, Ontario.

"If nothing else, Ted's account has been a great training ground," said Robert Korthals, former president of the Toronto Dominion Bank, which first lent money to Rogers in the early 1970s. "It never sleeps. It's fun to bank people who do things." And profitable, too, especially if the borrower pays his bills as Rogers does.

To maintain his below-investment-grade debt rating, Rogers continues to pile on the debt at a prodigious but not unbalanced rate. His shareholders, meanwhile, have not received a dividend in ten years.

But they're not complaining. A $100 investment in Rogers in 1984 is today worth more than $1,300.

All this began with that initial investment of $85,000 for a radio station with an audience of old men in cardigans. In building an organization that reaches one in four Canadians, with current revenues of more than $2.2 billion, Rogers has seldom been out of debt. While his company has shown a profit only infrequently, Rogers has mortgaged his house, tapped his friends and come close on several occasions to losing it all. Like a shipbuilder using a rowboat as collateral to build the *Titanic*, he has borrowed continually to finance his dreams relying on his own reputation and the promise of his companies to generate wads of dough.

When he started in the 1960s, $85,000 was not an insubstantial sum. For $16,000 you could buy a three-bedroom suburban bungalow in Etobicoke that would now cost about $200,000. For $85,000, you could buy a mansion in Rosedale. It was certainly more than pocket change for a law student like Rogers.

And it came in handy, when Rogers received his first cable licence from the CRTC in 1966, that the Bassetts and the Eatons took a 50 per cent interest in Rogers Cable while providing two-thirds of the financing. But Rogers has moved far beyond the category of a rich kid with well-heeled friends and a privileged background. Nor has he succeeded at the entrepreneurial equivalent of shooting fish in a regulatory barrel. Other cable companies such as Standard Broadcasting have actually lost money in their cable adventures. Meanwhile, Rogers learned early that financing comes at a cost, personal as well as financial. When it comes to business, even his friends expect to get their money back.

In 1970, when the CRTC prohibited cross-ownership of TV stations and the cable companies that distribute their signals, Rogers made a new friend – the Toronto Dominion Bank. Under the new regulation, the Bassetts and the Eatons had to repurchase Rogers's share of their broadcasting company, a challenge they had no trouble meeting. Rogers, however, had a little more trouble raising the $2.5 million he needed to buy back his friends' share of his cable company.

"The Eatons and the Bassetts were very gracious," he said, "but I was in a lot of trouble. I acquired their stock with no money down and eighteen months to pay. The eighteen months came and went, and I didn't have the money, so it was extended for a number of three-month periods. Finally, on July 1, 1971, I tried to sell a debenture issue, but the government had started a capital-gains tax, and everyone was very leery of investing."

This was in the good old days before Rogers and his partners in Unitel could afford to lose $1 million a day. In fact, the loss of $1 million in 1971 would have put him out of business. Desperate to avoid a catastrophe, Rogers looked for every way possible to save money. With this in mind, Rogers's top dog, Phil Lind, asked one of his neighbours, James Thackray, for a favour. Lind had worked on political campaigns with Thackray, who, in real life, was head of Bell Canada in Ontario. At the time, Bell had little interest in Rogers and his enterprises. "[Thackray] told me once if I ever needed a favour, I should ask," Lind said. "So I asked him if we could hold off paying our monthly pole charges for sixty days. Thackray said 'sure' and took care of it. We couldn't even meet our payroll. He could have wiped us out, and it would have been a very different story."

With a deadline less than twenty-four hours away, Rogers and his wife finally raised the $2.5 million they needed through Unas Investments Ltd., signing the papers just before midnight and taking a second mortgage on the family house, which had once belonged to another media pooh-bah with less devotion than Rogers to Canada,

Jack Kent Cooke. Years later, Rogers still credits TD for connecting him to Unas. "Dick Thomson [chairman and CEO of TD] saved this company," Rogers said. "That was one of a number of close calls." More than twenty years later, Rogers would also repay his neighbours' generosity by investing $13 million in Baton Broadcasting Inc.

Having weathered the storm, Rogers began his expansion in earnest. Borrowing against cash flow, he acquired small systems throughout the country. But he had his eye on bigger baubles for his cable chain. In 1973, he made his first approach to Canadian Cablesystems Ltd. in Montreal. CCL had twice as many subscribers as Rogers and wasn't about to let this rosy-cheeked Upper Canadian step into its territory.

Rebuffed but determined, Rogers continued to build his own company, relying heavily on loans. "We were able to borrow a much higher multiple of cash flow," Rogers said. "For the banks, our business, although capital intensive, actually had far less risk than manufacturing or minerals."

In 1978, Rogers was back on CCL's doorstep. This time, CCL had to listen. With an $18-million loan and the cooperation of Peter and Edward Bronfman, Rogers finally succeeded in buying CCL, creating in the process the largest cable company in Canada, with $58 million in revenue in a $235-million industry. In effect, the acquisition was a reverse takeover. Having acquired CCL, a public company, Rogers then incorporated his own cable, pay-TV and converter businesses into it, in return for $34.6 million. The company proceeded under the name CCL until 1981, when Rogers changed its name to Rogers Cablesystems Inc.

By then, Rogers had become the largest cable operator in North America. At one point, his company's subscription list surpassed 2.5 million households, which made it the largest cable operation in the world. For the first time, Rogers Cablesystems ranked among the top 500 companies in Canada, with sales of $109 million and assets of $421 million. In reaching this point, its debt had increased by fifty-two times over the previous five years. With the acquisition, for US$152 million, of UA-Columbia Cablevision, the ninth-largest cable company in the United States, long-term debt reached $164 million from just $15 million in 1979. In the same period, capital expenditures had risen to $180 million, up from just $10 million, while net income fell to $3.4 million from $10 million. In fact, it would be one of the last times that shareholders would see any profit at all from Rogers.

But as Rogers continually pointed out, cable companies operate like a utility. "The banks are encouraged by our utility-base earnings situation, which gives steady cash flow and offers excellent growth potential."

"The loss you take on the first few subscribers you sign up is enough to blow your ears off," added Robert Francis, then Rogers chief financial officer and a ten-year veteran of the expansion battles. "But the solid base of assets you're creating ensures that there will be a pot at the end of the rainbow."

Added Colin Watson: "The beautiful thing about cable is that it's like a utility, only with an upside, because you have to be incredibly inept to lose customers. That's why you can't really drive a cable company over the edge." Unless of course you operate the company in Canada, where government regulations limit rate increases to 6 per cent at the same time as wages are rising by 17 per cent, as they did at Rogers in the early 1980s.

At the time, interest rates were rising through the roof as Canada floundered once again in another of the many recessions that seem to descend every few years like a plague of financial locusts. Even the banks became edgy about Rogers, as the recession threatened to swallow the company, pot of gold and all. They urged him to get his balance sheet in order, and they told him to sell assets and adopt a strict financial plan. "The bankers said, 'Let's get a program,' and we did. No complaints. We'll get on with it," Rogers remarked.

Others involved at the time took a less jaunty view of the situation. "We were hung out a mile," recalled Watson, "losing money like mad, and with a huge bank debt. All our friends and neighbours began feeling sorry for us. The company was cash poor for quite a period."

Following the bankers' orders, Rogers sold Famous Players in 1981. He had acquired the company when he bought CCL. In fact, Famous Players had been the original name of CCL before 1971, and was generating more than half of Rogers Cablesystems's profits. Rogers sold it for cash and shares worth a total of $36 million.

Still in need of cash, Rogers approached United Artists Theater Circuit, his U.S. cable partner in San Francisco, to buy Rogers's existing U.S. assets in the name of the partnership. Without the burden of its U.S. holdings, which were gorging on capital and wouldn't break even for another six years, the deal would free up Rogers to borrow more money to finance his expansion. But UA Theater Circuit refused.

Burdened with floating-rate debt, facing imminent bankruptcy, Rogers was once again in deep doo-doo. "We sold off everything we could just to keep afloat," he said. "It was like flying a plane and you're tossing stuff out just trying to keep above the trees." Rogers managed to keep the plane flying long enough to carry him to Los Angeles, where he dropped in on Michael Milken at Drexel Burnham Lambert Inc. in Beverly Hills.

In his relatively brief career with Drexel, Milken raised some $26 billion for such companies as MCI, McCaw, Viacom, TCI, Time Warner, Turner, Cablevision Systems, News Corp. and a host of other cable, telecom, wireless and entertainment companies. As U.S. writer George Gilder points out, "None of these firms commanded substantial collateral acceptable to a bank, and thus they could have raised these billions nowhere else." Today, these companies are collectively worth some $224 billion and are the foundations of a national information infrastructure unrivalled in the world. For his contribution, Milken went to jail.

But not before he helped Rogers. With Graham Savage, then Rogers's vice-president of investments, Rogers approached Milken in Drexel's trading room to ask for $300 million. "There was Mike sitting before this huge bank of telephones," Savage recalled. "While we were there, he literally raised more than $1 billion in 30 minutes." Milken also raised the $300 million that Rogers needed, which Rogers used to pay down his bank debt. "That issue really saved our bacon," Savage said.

With his bacon intact, Rogers came back to life. "It was like watching the Indian rope trick," said an associate. "One minute, Ted was standing there with nothing but a bunch of hocked properties in his pocket. The next he was climbing up the rope so fast he was practically out of sight."

Still, with interest rates over 20 per cent, and with his cable operations expanding through the United States, from Syracuse, New York, to San Antonio, Texas, Rogers's board was in no mood for more adventures. After a day of debate in 1982, his directors rejected Rogers's idea of investing in a national cellular system. Heck, the company could hardly raise a dime for a phone call. So Rogers invested his own money, in partnership with Marc Belzberg and Philippe de Gaspé Beaubien. By 1990, Cantel was worth $5 billion, based on the private market value of its assets, and by then it too belonged to Rogers Communications.

The board also rejected Rogers's idea of expanding the company's Irish cable system into Europe. Rogers was shocked. "I looked around the boardroom table and saw every single person was against it," he said. "I could have made it fourteen votes against one controlling vote, but I knew in my heart of hearts that they were right. I'd come to realize that we are a fragile company compared with other world firms. We can't sustain big write-offs."

Rather than expanding the Irish system, in fact, Rogers sold it the next year, for $5.7 million. He also sold his interest in the Vancouver Canucks hockey team, a cable system in Syracuse and a one-third interest in Cableshare of London, Ontario, raising more than $24 million.

As his board knew, not all of Rogers's ideas turn out so well as Cantel. In 1983, for example, Rogers invested more than $15 million in new fixed assets for a pay-TV system. In its first year, the system lost $12 million after interest and depreciation. Rogers had spent several million to promote pay-TV. He also paid $85 for each descrambler, another $30 to $50 on sales commission, installation and disconnection costs per subscriber. For his efforts, he said, he received $7 a month for two or three months, until most subscribers decided they didn't want pay-TV at all. The investment didn't produce a single cent in operating profit, prompting Rogers to state, "We cannot continue to make investments on this basis."

By 1986, Rogers Cablesystems had borrowed a total of $750 million to finance its growth and had lost money for five consecutive years. Even in the worst of times, Rogers insisted on spending more money than any other cable company on research and development and capital equipment. A computerized billing and order-taking system, for example, installed in 1982 at a cost of $7 million, eliminated the need for a personal visit from a Rogers rep, at a cost of $20 or more per visit, to connect or unhook a subscriber. The company's ratio of debt to equity stood at 6:1. But for Rogers, the pursuit of profit is not nearly so interesting as seizing opportunities and trooping to Ottawa with his band of happy warriors to defend them from predators.

Still, Rogers knew what he had to do, and he began once again to unload assets, raising $300 million by selling some of his U.S. holdings and reducing his cable subscriber base to 1.8 million in southern Ontario, Vancouver, Portland, Minneapolis and San Antonio, Texas.

Rogers had also started to reduce the level of foreign ownership in his company, which he knew would further his broadcasting ambitions in Canada. Under regulations passed during the reign of Pierre Trudeau, before a company can hold a broadcasting licence in Canada, Canadian citizens must own at least 80 per cent of its voting shares and 80 per cent of its paid-up capital. In 1986, foreign owners held more than half of Rogers Communications's non-voting shares. To reduce their numbers, Rogers de-listed his company's shares in the United States. In May 1987, the company also prohibited the transfer of its voting and preferred shares to non-Canadians. Later that year, Rogers offered $25 a share to buy back up to 10 million non-voting shares. He advertised the offer in the *Wall Street Journal*, but not in the Canadian business press.

The move had the unintended effect of drawing attention to Rogers Communications stock. Writing in *Barron's* weekly in March 1988, U.S. money manager Mario Gabelli said that Rogers stock was undervalued. Rather than $25 a share, he said, it was worth closer to $100. He also forecast that it would rise to $133 a share by 1989 and $202 by 1991.

It took a U.S. commentator like Gabelli to convince investors that Rogers was worth looking at. Shares in his company rose, as predicted, from $25 to $96 by April 1989 and to over $100 by June. And many of the shareholders who saw their wealth quadruple in the process worked for Rogers. "There are lots of middle managers around here who have accumulated real wealth in this company in the last few years," said James Sward.

Despite Gabelli's pronouncements, Rogers succeeded in buying back almost one-third of the outstanding shares in his company. By the end of 1988, non-Canadians held fewer than 5 million shares, not the 20 million of a year earlier. In the process, Rogers had reduced foreign ownership to 35 per cent from 75 per cent and reduced the public float of Rogers stock by 30 per cent. Meanwhile, he had also reduced his company's debt from more than $900 million to under $350 million, in part by selling his U.S. cable assets to Houston Industries for $1.6 billion.

For a brief moment, in fact, the company was almost debt-free, the financial equivalent for Rogers of yogic flying and with as much permanence. Soon Rogers was plunging back into the debt pool. As the paper value of his company surpassed $3 billion on news of the U.S. sales and share buy-back, Rogers promised to borrow more than $1 billion to expand his operations and improve the quality of their service. In keeping with the company's characteristic R&D and capital spending patterns, Rogers said he would spend $525 million over three years to refurbish his cable operations, replacing coaxial cable with fibre-optic networks to deliver better-quality pictures and eventually to accommodate home banking, pay-per-view and other gewgaws of the telecommunications age. He also said he intended to spend $600 million on Cantel over the same period. The money would come not from accumulated cash but from the company's $1.5 billion in bank credit. Debt again, said investors.

Throughout 1989, Rogers expanded his Canadian holdings, acquiring the systems of Western Cablevision Ltd. in Vancouver and

Astral Bellevue Pathé in Montreal. With Astral Bellevue Pathé, Rogers gained a foothold in First Choice and Premier Choix, the company's pay-TV operations. Meanwhile, through Western International Communications, he bought a share of Canadian Satellite Communications Inc., whose satellite services are used to direct TV signals to remote locations and would later be used for direct-to-home satellite broadcasting in competition with cable. Cancom also provided Rogers with another piece in his long-distance puzzle. He could use the satellite system to distribute wireless signals among his Cantel subscribers across the country.

Having practised a few steps in the acquisition cha-cha over the first half of the year, Rogers was ready for his big dance. In September, he acquired 40 per cent of CNCP Telecommunications for $275 million, again using borrowed money. To build a system that could compete against the phone companies, CNCP would have to invest up to $1.5 billion in capital costs and would not turn a profit until 1999. (So far, it's right on schedule, spending money like crazy without making a dime.)

By then, Rogers had so many irons in the telecommunications fire that one of them was bound to get hot. At the time, for example, Cantel contributed only 5 per cent to Rogers operating profit. Within five years, that figure would rise to 40 per cent. Before that happened, however, Rogers was sorely tested again. In August 1990, with the economy fluttering into another downward spiral, Rogers came within eight hours of financial ruin. The company had been buying back its Cantel service centres from Cantel's dealers, spending more than $30 million in the process. It had also expanded its interest in MTV, purchased the assets of a cable system in Anchorage, Alaska, and assumed complete control of Western Cablevision. Life was never boring at Rogers. But as interest rates flew upward, Rogers had to pay the piper. "It was like going down a coal chute," he said.

Canadian Pacific, his partner in Unitel, helped out until he could replace $3 billion in bank debt by selling bonds and shares, primarily in Cantel. Meanwhile, Rogers sold his investment in Western International Communications, but not in Cancom's satellite operations. "It seems to happen to us about once every ten years," he said.

It might happen less frequently if Rogers were a U.S. company. Canadian ownership regulations to the contrary, Canadians have never embraced Rogers wholeheartedly. Despite his astonishing success and the wealth he has created for his shareholders and employees, most of whom are Canadian, Rogers's fellow countrymen have been less inclined to invest in his company than Americans. "Why don't Canadians buy their own god-darned companies," he asked in frustration in 1989. "Why do they buy bonds and things like that?"

With his penchant for qualified risk and his inclination to bet the shop on a new venture, Rogers appeals to risk-takers, not to the superannuated United Empire Loyalists collecting pensions from the library board who seem to control much of the investment money in Canada. "American investors have always liked Rogers," said Toronto investment manager Ira Gluskin. "Canadians are upside down." While Canadians worry about what Rogers will do with all his loose cash and borrowing power, Gluskin observed in the *Financial Times*, Americans look at his historical rate of return on investment and calculate how many billions more he can make on his initial stake. "Only tough, restrictive, discriminatory, unfair legislation has kept this company from being totally owned by foreigners, outside of the block held by Rogers and a few others," Gluskin observed. "To maintain his standing as a Canadian communications company, he has actually had to discourage American investors – the only group that believed in him." Through his own investment company, Gluskin has put his money where his mouth is. "We've made more money in Rogers than in any other Canadian stock," he said in 1994.

Rogers's entrepreneurial zeal and his penchant for relentless deal-making have occasionally carried him close to the edge of the regulatory precipice. The investment community didn't know for several weeks, for example, that he had suffered a coronary aneurysm in the 1980s, and Rogers said he hadn't thought to tell anyone. But for a company that depends so heavily on the vision of one man, the information would certainly have affected the price of shares in his company.

Around the same time, regulators did take notice of Rogers, when he began buying shares privately in Moffatt Communications Ltd. in Winnipeg, then offered to raise his personal stake, in April 1988, to 21 per cent from 9.3 per cent. As required, he submitted an application to the CRTC for permission. But Randall Moffatt, controlling shareholder of the company bearing his family's name, did not disclose Rogers's offer to the public. As time passed, Rogers's plan became known first to associates within each company, then to CRTC courtiers, pencil-pushers and party-goers in Ottawa, where gossip passes for intellectual discussion and not everyone abides by the high standards required of public office. By August, Rogers's offer was probably the talk of every bun fest and political weenie roast in Ottawa, Winnipeg and Forest Hill. Not surprisingly, as every matron with money in her pillowcase ordered her broker to buy Moffatt shares, the price of Moffatt rose to $23 from $14 between April and August.

Moffatt, however, said confidentiality prevented him from disclosing the offer, even on a confidential basis, to officials of the Toronto Stock Exchange, who could have monitored trading to see who was cashing in on the price increase. Meanwhile, CRTC regulations, written in the bureaucratic equivalent of Swahili, had been changed in 1986, placing the responsibility with licencees like Moffatt to inform the CRTC of Rogers's offer. But the new rules didn't apply to Moffatt, because his company held its broadcasting licence before the new rules came into effect. In the midst of the hubbub, Rogers simply changed his mind and abandoned his offer.

A year later, though, as he prepared to buy eleven radio stations from Maclean Hunter, he sold his 9.3 per cent interest in Moffatt – 480,000 shares – for $25.37 apiece. Rogers claimed that he didn't want any conflicts to arise between his application for the radio stations and his holdings in Moffatt. The media swallowed the explanation, and Rogers swallowed a profit of $5.5 million.

Another minor brouhaha erupted in January 1992, when Rogers sold 2 million non-voting shares of his company at $14.50 each to ScotiaMcLeod Inc., with whom he had maintained a relationship for more than ten years. Rogers said he was getting his estate in order. As required by regulators, he had announced his plan the previous November, when he sold an initial 4 million shares to ScotiaMcLeod, aiming to reduce his own personal debt of $75 million. "I'm fifty-eight," he said, "and being out of debt personally is something that I think is very important."

The January sale came on a Thursday. The next day, Rogers left town for New York. That afternoon, at 4:45, after the markets had closed, a suspendered honcho from Gordon Capital Corp. descended a few floors in the Toronto Dominion Tower to Rogers's corporate offices to propose a deal to Graham Savage. Gordon would buy 13.3 million Class B shares and 6.2 million warrants from Rogers Communications in one fell swoop, for just over $200 million. Then Gordon would turn around and resell the shares at a profit.

Savage said the call came out of the blue. On Saturday morning, he phoned Rogers to tell him about it. Over the rest of the weekend, Rogers and Savage conferred with the company's directors and obtained their approval for the deal. Meanwhile, three other investment dealers joined the party: Midland Walwyn, Wood Gundy and, lo and behold, ScotiaMcLeod. Rogers went for the deal, and the company made an announcement on Monday. As is customary after a new share issue, the value of Rogers shares immediately fell by $1.13 to $13.87.

Critics suggested that Rogers had deliberately dumped his own stock two days in advance of the deal with Gordon. "The feedback I'm getting is that some people are pretty pissed off," said one analyst. Rogers must have known about the Gordon deal, they claimed, before he sold his own shares. "If you believe that a company like [Rogers] does that kind of financing in the time frame described, you probably also believe the earth is flat," said an anonymous lawyer. "There are exceptions, but, as a rule, to put together an issue and price it and do the legal work takes months, especially in markets like this."

Others gave Rogers the benefit of the doubt, but criticized him and ScotiaMcLeod for their clumsiness. "If Ted Rogers erred here," said another analyst, "it's the same error made by the underwriters by not realizing what the reaction would be to the way this looks." Another said the deal "looked so stupid it lends credibility to the idea that it was a quick deal."

Rogers insisted that he knew nothing about Gordon's impending deal when he sold his own shares on Thursday. The first he heard of the deal came on Saturday morning, he said, when Savage called him in New York. Even the close-mouthed head of Gordon Capital, James Connacher, whose controversial financial manoeuvrings had attracted so much media attention that he finally stopped talking in public, condescended to make a public statement. "It may look odd, but there was never any indication from Ted Rogers or the company that they wanted to do anything until Sunday afternoon," he said, adding that until the deal closed on Monday morning, he had not talked to Rogers for two months.

Rogers said critics of the deal were out to lunch, that he often holds board meetings at short notice, and that his company, which is constantly in the process of raising money, has boiler-plate documentation for such deals on the shelf. ScotiaMcLeod itself had put together a $200-million issue of convertible debentures in 1990 in just two days, Savage added.

Regardless of how it looked, Rogers said, he wasn't about to turn down an opportunity to raise $200 million for his company. "Besides," he asked, "how many other Canadian companies can raise $200 million over a bloody weekend?" The Toronto Stock Exchange and the Ontario Securities Commission both sniffed at the deal, but soon scampered away to turn over more promising rocks in the market garden.

In pursuing his estate plan, Rogers later sold another 7.2 million shares in 1993, worth about $148 million. "He's sixty," said Graham Savage, "and no one at that age wants their entire net worth wrapped up in one company."

When Rogers held his annual meeting in April 1992, in Montreal and Vancouver simultaneously, through the wonder of fibre optics, he announced that his company had reduced its debt by $200 million, to a mere $1.8 billion. The number of subscribers to cable and cellular services had risen, and revenues exceeded $1 billion. Rogers was still spending twice as much as its operating profit to install the equipment it needed to serve its subscribers. And the company still wasn't making a profit. But what the hey, you have to spend money to make money, Rogers pointed out, even if it takes twenty years. An audacious soul in the audience asked when Rogers might eke out a profit. "Probably in the first year of my successor," Rogers quipped. Meanwhile, shareholders approved the creation of almost 1 billion non-voting shares in the company.

Naturally, they were non-voting shares. Rogers has relied for years on the typically Canadian practice of issuing voting and non-voting shares to raise equity capital. While he holds the vast majority of voting shares, he holds only 4 per cent of the non-voting shares. This gives him control of the company without ownership. It's a controversial though common practice in Canada. In fact, about 80 per cent of the firms listed on the TSE have a controlling shareholder.

Among other things, the arrangement presents a challenge to directors of companies like Rogers, because there's little distinction between management and control. "As CEO and chairman of the board, controlling shareholders are very much in charge," observes Professor James Gillies of York University in Toronto and author of *Boardroom Renaissance: Power, Morality and Performance in the Modern Corporation.* "But they also issue stock to public shareholders. So their board members owe a duty of care to the corporation while maintaining a fiduciary responsibility under the law to the shareholders.

"The board should have something to say about the corporation's strategic direction, management performance and other issues," Gillies adds. "But they can't do their job if the controlling shareholder doesn't want them to. If the controlling shareholder wants to take the corporation in a direction that the directors don't like, their only option is to resign."

The distinction between voting and non-voting shares also influences shareholders' decisions to buy stock in Rogers. In the case of an offer to buy Rogers Communications, for example, only holders of voting shares would participate. Since 1987, the TSE has required issuers of new non-voting shares to include a coat-tail provision that ensures their purchasers that they will participate in the wealth of a takeover bid. Rogers shares, issued long before 1987, carry no such provision. Instead, holders of Rogers non-voting B shares receive a dividend; holders of A shares don't.

Rogers makes no apologies for the practice. "I've been very straightforward," he said. "You can buy the A's, or you can buy the B's and get the dividend. Don't come crying to me if there's an offer later for the A's."

Of course, individuals don't invest in Rogers for the dividend. They invest for the growth. And in this respect, long-term shareholders have little to complain about. Between October 1987 and July 1988, for example, Rogers B shares soared from 14½ to 53, fueled by his plans to sell his U.S. cable holdings. Cantel shares have provided equally impressive returns, rising at one point from $58 to $92 in a single eight-month period.

As Steven Garmaise of First Marathon Securities said in 1989, "If you're looking for an earnings-driven, dividend-paying company with a conservative balance sheet, don't buy Rogers. But if you're looking for a company that has a different way of enhancing shareholder value by building up very attractive assets and buying into things very cheaply, this is clearly one of the places you want to be."

Rogers himself has said on several occasions that profits don't really interest him. "Worrying about next year's profits is soul-destroying," he said. "All this company has to do to make money is stop growing."

Losing money has other advantages. In the process of recording losses on the balance sheet while making equally enormous capital expenditures, Rogers has accumulated a closet full of tax concessions. By 1994, they were worth more than $500 million, and Rogers would soon put them to good use.

Contrary to the paranoid impressions of his more rabid but less-informed critics, Rogers has never simply pocketed his own earnings. He has consistently invested more in his company than the rest of the industry and far more than many entrepreneurs who simply want to finance their club memberships and vacation home. "Rogers has never shied away from making the necessary capital investments to remain a leader," said Graham Savage. "Over the last five years, our capital expenditures have averaged about $400 million annually. With competitive pressure escalating, we expect to increase capital spending at [Rogers] Cablesystems over the next few years. Our focus will be to continue to upgrade our plant and those of our newly acquired systems, so we can deliver new products and services.

"Net profits are less important to us, because they're after non-cash items such as depreciation and amortization, which are high in a business like ours that demands continuing capital expenditures and has grown in part through acquisition. However, net income is likely to come hand in hand with the realization of free cash flow.

"We recognize that in order to expand our shareholder base, we need to produce more traditional financial results. So we anticipate seeing free cash flow and profits sooner rather than later."

But investors didn't start holding their breath. With $2 billion in untapped bank credit, Rogers had lots of ammunition to fight more battles for growth, and to heck with the darned profits. Some reporters speculated in 1993 that he would increase his investment in CNCP. Others suggested that he would take his company private, speculation that Rogers had denied for years and that made no sense anyway, because it would have prevented him from taking the company public again without tying non-voting to voting shares, a prospect that gave him an attack of the financial chilblains.

Instead, he surprised them all when he made an offer of $17 a share for Maclean Hunter. He had already purchased a large block of the company's stock through First Marathon. Now he intended to buy the whole company. Compared to his initial efforts to raise money for his company, putting together $2 billion in bank lines was as easy as turning on a TV. "I went to see Peter Godsoe, head of the Bank of Nova Scotia, on a Monday night," he recalled, "and by that Friday we pretty well had the $2 billion." As part of the deal, Rogers moved his corporate headquarters across the street to Scotia Plaza. In addition to his bank lines, Rogers had $750 million in cash and $65 million of his own money. So much for estate planning.

By now, Maclean Hunter's hundred-year-old magazine publishing operation, upon which the company had built its name, still contributed 40 per cent of its revenues. But they were technological dinosaurs, contributing only 4 per cent of the company's operating profit. Cable was now the big Kahuna. Cable contributed only 26 per cent of the company's $1.74 billion in 1993 revenues; but the company derived more than 80 per cent of its operating profit of $212.5 million from cable operations.

Throughout the bargaining, reporters and analysts criticized Maclean Hunter for its cautious and conservative approach to the information age. They questioned the company's reluctance to take risks, its emphasis on dividends rather than capital spending, its aversion to debt and its maintenance of operations that contributed little to the company's bottom line. Some of this was warranted. A technology-follower, Maclean Hunter took several years to recognize the value of a computer. Long after most journalists had abandoned their typewriters for PCs, reporters at the *Financial Post* and *Maclean's* were still pounding away at their Underwoods, then cutting and pasting their manuscripts and submitting them for retyping into a mainframe by a team of underpaid workers in a windowless back room on the fifth floor of the company's downtown office building. When the company finally installed a bunch of Zenith computers in its business periodicals division, they were obsolete before anyone even plugged them in.

If Maclean Hunter had been timid and cautious, it had also been consistently profitable, paying a dividend every year for decades. Its shareholders had come to expect as much from their company. In defense of Maclean Hunter's record, a frustrated Ron Osborne said, "Does anybody seriously believe the Street would have allowed us to do Cantel, to incur the kinds of losses, to make the kinds of investments required to take a ten-year horizon to develop a whole new business without a single earnings-per-share benefit over the entire ten-year period? We would have been cut up and dissected and broken apart ten years ago had we gone into that."

Maclean Hunter had been involved in new multimedia projects, he continued, but it had not committed large amounts of capital. "We have experimented left, right, and centre," he said, "but on such a small scale that the company's income statement could absorb the cost every year without somebody from the Caisse de dépot et placement du Québec or the Canadian National pension plan worrying that we have truly hammered the profit and loss statement, and that they are going to be worried about EPS growth for the next five years."

As part of the Rogers empire, the company had now acquired an entrepreneurial leader and a junk-bond rating. If it came down to a choice, however, the leader would be worth far more than the rating. Rogers himself had set an unsurpassed record of focused growth. Profitable or not, his company had provided shareholders with a far greater return over the last twenty years than Maclean Hunter, partly because the man who owns the company also runs it. As a study by the Wyatt Co. in the United States has shown, a company whose chief executives own a large stake of its shares performs far better than a company with hired guns at the top. "Maclean Hunter has been superbly run, probably better run than our company," Rogers said graciously in defence of his latest victim. "But transferring these systems to Rogers will mean that they will be restored to entrepreneurial leadership."

Rogers immediately started to put his tax-loss carry-forwards to good use. The Toronto Sun applied $8.9 million of losses to its own taxable income, producing a profit of $1.2 million, which it wouldn't have shown otherwise.

Since the acquisition, Rogers shares have fallen in value, to less than $14 from a high of $23. Rogers has lost $1.6 billion in market capitalization in the process. As regulations opened the door to competition, as Unitel lost money at the rate of $1 million a day, and as Rogers struggled with a huge debt load, investors grew wary of the company. Yet despite its reputation as a debt-laden, profit-averse company, Rogers's debt-to-cash-flow ratio of about 4:1 is better than the comparable average ratio in the United States of 8.4:1. At Rogers Cablesystems, debt-to-cash-flow is 4:1, and the company has grown ten times larger than it was in 1985. Revenue in 1994 rose 24 per cent, and cash flow rose 99 per cent. Meanwhile, Unitel has spent heavily on capital equipment, as it promised to do in 1992, while chopping staff from 3,700 to 2,750. If it stopped spending on capital improvements now, it would make money, according to president and CEO Stan Kabala. Contrary to the fears of some, Unitel's losses have no immediate impact on Rogers Communications. "Our businesses have always operated separately with their own managements, boards of directors and separate financing arrangements," said Rogers. "These will continue, but much more focus will be put on leveraging opportunities for common action and seeking economies of scale and scope."

With this in mind, a number of analysts recommend Rogers as a growth stock. "RCI's existing investment in Unitel is already effectively written off in its current share price," points out Jeremy Burge, an analyst with Toronto Dominion Securities. "Even if Rogers purchases CP's 48 per cent stake in Unitel with 11.1 million RCI B shares, Rogers remains undervalued."

The bigger his company gets, however, the more it comes under the scrutiny of armchair presidents and weekend financial geniuses. They say that Rogers is a one-man show with a junk-bond rating and a 4:1 debt/equity ratio that keeps it constantly on the brink of disaster. Vulnerable to competition from all sides, Rogers's company, they say, is worth nothing more than the wires in the ground, unless it starts to make a profit. They conveniently ignore the fact that Rogers met a $2.1-billion obligation within months of acquiring Maclean Hunter; that if the cable operation is temporarily spinning its wheels, Cantel is doing fine; and that, beneath the ugly mug of Unitel lies the heart of a profitable business. The days of guaranteed expansion in cable may be over. But the days of consolidation and open competition in telecommunications lie ahead, and they promise to bring out the best in Rogers.

CHAPTER 8

The Man

Who is this guy?

May 27, 1933 (Year of the Rooster): You have enormous drive.
You are very self-confident, but should also allow others to
have their say. You are happy, optimistic and warm-hearted.
You keep yourself busy and are rarely troubled by trivialities.
Occasionally you quarrel unnecessarily with your friends,
and it is important for you to learn to control your words. You
are particularly suited to a career as a sole owner or
proprietor.

KWOK MAN-HO
Chinese Horoscopes Library

Canada is a nation where several years may pass before our telecommunications regulators see what's happening in the industry they're
supposed to regulate; where bureaucrats earn more respect than the
unseemly and aggressive entrepreneurs who pay their salaries; where
an unqualified patriot faces the daily risk of a broken heart; where
entire provinces bow down and pay homage to the paternalistic families of potato farmers, grocers and oil magnates who dominate their
economies; and where anyone who tries to rise above his station

receives a rude awakening. In such a nation, it seems only fitting that our most accomplished telecommunications entrepreneur would be a fatherless blueblood with a law degree, one good eye and a reconditioned heart.

In May 1972, Ted Rogers turned thirty-eight, the age at which his father had died. If he took the time to reflect on his life so far, he might have made a few comparisons. At thirty-eight, his father had developed a new type of radio receiver, then started a radio station so people could use their receivers to listen to something more entertaining than cosmic noise. He had pushed back the frontiers of radio, dabbled in TV, then put his mind to radar before he died.

At the same age, his son now owned three radio stations, had started a fourth, had co-founded CFTO, one of the first commercial TV stations in Canada, and had taken his first steps into the cable industry with Rogers Cablesystems. Unlike his father, whose death had left his mother strapped for money, Ted Rogers had already developed an estate plan for his young family. And while he hadn't revolutionized the world of communications, he had given Toronto an alternative to Walter Bowles, Ed Fitkin and Uncle Bing Whittaker on AM radio. His father might have performed more impressive technological acrobatics in his short life, but Ted Rogers could juggle with the best of them, and he still had a long way to go.

Only twelve years had passed since he had started Rogers Radio Broadcasting Limited and acquired CHFI, a semi-dormant FM outlet in Toronto. At the time, he was finishing his law studies at Osgoode Hall, struggling with torts and the theory of uses in property class and booking boogie-woogie dance bands to earn money and keep his head from exploding with the mumbo-jumbo of a lawyer, which he really didn't want to be anyway. While articling with Tory Tory DesLauriers & Binnington, he had arranged to broadcast CHFI's programming over an AM channel. For that purpose, he bought CHFI-AM, which was later renamed CFTR. By 1962, when he was called to the bar, Rogers had also started a TV station with his friends. He was twenty-eight. From then on, he never gave another thought to making partner in a law firm.

Over the next thirty-four years he would become a husband, a father and a wealthy man, as well as a legend, a czar, a king, a mogul, a whiz and a wild man in his industry. On occasion, he would come close to losing it all, using his company, his personal holdings and the family house as collateral to pay for one more episode in his story. At sixty-one, an age when many men start to consolidate their acquisitions, get their affairs in order, buy a little condo in Fort Myers and retire to the practice tee to whack a few balls into a field, Rogers was still racing around town in a suit trying to convince bankers, investors, lawyers and regulators that he hadn't chugged too many mai tais when he said he wanted to add another company to his holdings, and, "Gee whiz, I can get it for only $3 billion and change."

By then, the financial community had seen enough of Rogers to know that he wasn't blowing smoke. They also knew that he wouldn't take the money and buy another hideaway in Nassau. In fact, money for Rogers has always been a tool, not a goal in itself. "I'm never going to run my company only for money," he once said as he explained for the 489th time why his company seldom turns a profit. "I won't do anything foolish, but I also won't hesitate to spend money on things that will propel Rogers Communications Inc. into the future."

Even his wife, Loretta, chides him for spending more than any other cable company in Canada. But to Rogers, such spending keeps his company in the communications race. "You destroy yourself if you're not first class in everything you do," he said in 1989, after pledging to spend $525 million to upgrade his cable system. He even increased the budget at his little community cable station by 300 per cent, so that the seven viewers of the channel wouldn't wonder why the mayor of East York had a purple glow around his head. "The operating people think I've lost my marbles," Rogers said. "But the phone companies or someone is going to try to overwire us at some point. We've got to have the best possible service so that nobody would dare take us on."

He pays himself and his executives fairly, and provides them with rich incentives to perform. But their salaries are not out of line with the rest of the industry. In fact, base salaries for Rogers executives have been frozen for three years. In 1994, his own salary was $485,000. That's the equivalent of a mediocre major-league baseball player. He also paid himself a bonus of $1.3 million after the acquisition of Maclean Hunter, putting himself in the same earnings category as a decent but not outstanding NHL goalie, although Kirk McLean probably has a more relaxing time in the off-season and doesn't employ over 13,000 people to prop him up in the net for the Vancouver Canucks. After the acquisition, Rogers laid off a total of 135 people; for each employee who remains – paying taxes, sending the kids to school and having their teeth fixed on the company dental plan – Rogers pays himself about $37. It's safe to assume that most of them would gladly pay that amount to keep their jobs. By comparison, BCE's chief executive, Red Wilson, paid himself $802,000 in salary and $590,000 in bonus in 1994, while Bell Canada, a BCE affiliate, will chop 10,000 people from its payroll between 1995 and 1997.

With his personal holdings in his own company, which have grown along with the business, Rogers can afford to live like a rich man. When he wanted to build a tennis court at his house in Forest Hill, he bought the house next door, which had belonged to Neil McKinnon, former chairman of CIBC, and tore half of it down. In addition to their pile in Toronto, Rogers and his wife own cozy little retreats in Muskoka and Nassau. In the Bahamas, he keeps a newly built 35-metre, $5-million yacht, the *Loretta Anne*, moored at his 2.5-acre estate, in view of his neighbours, who include the grandfather of mutual funds, John Templeton, and actor Sean Connery. To get there, he hops in the limo and rides to the airport, where he climbs aboard his private Challenger jet for a three-hour flight.

When he arrives, though, Rogers doesn't tear off his shirt and tie, grab a thermos of piña coladas and head for the beach. He just keeps on working. As one former executive put it, "Ted is absolutely brilliant within a narrow range, but he's not broadly based. He doesn't golf. He doesn't have hobbies. His company is his life."

They may not have his money, but gas-station attendants on their day off have more fun than Rogers. From Lyford Cay in the Bahamas, he stays in touch with his corporate lieutenants by fax and phone, working from a small office overlooking the ocean. Sometimes he brings executives and directors to his hideaway to discuss budgets and strategies. He maintains the same pace in Muskoka, dividing his day into segments of work and relaxation. But even in relaxation, he doesn't chit-chat about the Blue Jays' pitching problems or the lumps of chocolate in President's Choice Decadent ice cream. "Small talk doesn't interest him," said his stepfather and former chairman, John Graham.

At home in Toronto, a tie line keeps him connected to his downtown office. He seldom uses a computer, preferring instead to lug papers between his office and his house in five battered briefcases the size of laundry bags. When he gets up in the morning, he watches about seven minutes of news and business commentary in the bathroom before he heads off to work. "Actually, that's where I watch television the most," he said. Once in a while, he paces around in front of the TV screen as a movie like *Die Hard* or *The Terminator* competes for his attention with his own churning synapses. "I like exciting movies," he said, "movies with lots of action," although it's not easy to find a movie that's faster paced than Rogers's own brain.

Marriage didn't distract him much, nor did his four children, all born from 1968 to 1972, between trips to the bank. In the early days of their marriage, Loretta would drag him away from the business for a quick vacation, only to find that he'd stuffed his luggage with file folders, prospectuses and CRTC reports until there was no room left for a toothbrush. He might play a quick game of tennis or swim a few laps, but he'd phone the office as soon as he left the court or pulled himself out of the pool. In 1966, three years after their marriage, Loretta took him to Australia, where even Rogers couldn't make a phone call to the office to find out what his executives had done yesterday, because it was already tomorrow. Instead, Rogers read "an interesting little book" on a little-known technology. When he returned to Toronto, he started a cable company, which kept him tied up at the office for the next thirty years. Finally, his wife admitted that "His work is his hobby."

Hal Jackman, a high-school classmate and a lieutenant governor of Ontario, added succinctly: "Ted is not laid-back."

To be employed by such a man is an all-or-nothing proposition, as his employees well know. For a slide show prepared for a Harvard Business School banquet in his honour, some of them posed for a photograph on the lawn of their Don Mills headquarters, holding up two sets of giant letter cards. One set spelled out: "We love you Ted"; the other: "Most of the time."

A few former Rogers executives never loved him at all. "Ted is always charming in public," said one, "but, in private, if you disagree with him, his attitude is that he pays you very well, and that gives him the right to dominate your life and berate you."

Others take a more generous view. "He's such a nice guy that you don't want to get into a pissing match with him," said a broadcasting executive. "He's a passionate Canadian and filled with absolute confidence," said the former TD president Robert Korthals. "He goes away thinking we've reached a compromise that he's more comfortable with than we are. You feel bad afterward that maybe you've let someone get into too much trouble."

"Ted can be intimidating," said another former executive. "He can frighten me because he's so brilliant. But he's also off-the-wall. You think you can go into a meeting prepared, and you just have no idea what's coming next. There are times I get frustrated, times I'd like to kill him, but he sees things in a way no one else does. He's a risk-taker and he's a visionary."

Rogers is also a leader, not a manager. He moves from project to project, trying to focus on the big picture while understanding the details, maintaining control without interfering, and doing his best to delegate authority to his executives. He has the entrepreneur's ability to inspire his executives and light a fire under his deputies, and the entrepreneur's impulse to do everything himself. "You sit there and you listen," said a colleague in 1979, "and you ask yourself, 'What is this?' But let me tell you, you believe."

"This company runs on high octane," said Jim Sward in 1986. "I've been with it for seven years, and it's grown tremendously, particularly because of the intensity of the people who work here. And Ted gives them a lot of room. This generates an entrepreneurial spirit and sets a pace that's exhausting if you're not up to it."

Still, he likes to know what's going on in the company that he built. Little puffs of smoke shoot from his ears if an employee tries to pull the corporate wool over his eyes. "You can talk to me any day, but not to them, because I'm running this company and they're not," he said in 1987 to the *Globe and Mail's* John Partridge.

At a Cantel breakfast meeting in Montreal, he solicited suggestions from staff members for improving the company, writing them down as he walked around the dining room. Later he insisted that his managers follow up on every one.

"He likes to know everything that's going on," said George Fierheller, "rather than have decisions made by management committee or anything like that. He surrounds himself with people who can work in that environment. They're on their own to make recommendations, but they understand that they're going to have to get them cleared by him."

Fierheller speaks from personal experience. In 1988, he aroused the wrath of Rogers when he and other senior managers at Cantel sold some of their shares in the company to ScotiaMcLeod. Rogers had unexpectedly decided not to issue stock to the public but to keep Cantel a private company. The executives had expected to sell their shares on the open market, but they placed them with ScotiaMcLeod instead. The investment dealer then sold the shares to several institutions, which put them out of the reach of Rogers. Neither Fierheller nor the other managers had told Rogers about their plan, and he was upset, not only because he likes to know what's going on, but because he thinks key executives should hold shares in the companies they operate.

There's a fine distinction between control and interference, and, to the public, at least, Rogers seems to know where to draw the line. In fact, compared to other pompous media moguls, who seize every opportunity to hurl their multi-syllabic weight at the world through the pages of their organs, Rogers is shy and retiring. "The concept of Rogers sitting somewhere and phoning the radio news guy and telling him what editorial he'll run, then phoning the TV news guy and telling him what editorial he'll run, then phoning the cable company and telling them to kick CITY-TV off and put Channel 47 on, is as outrageous as it is impractical," observed Jim Sward before he left the company.

Rogers admits that his management style has its faults. "People like me are not well organized," he said in 1994 after acquiring Maclean Hunter. "I'm not being modest when I say that. I mean, look around," he added, gesturing towards his office strewn with the debris of a takeover's aftermath. File folders cover the tables and chairs. His sofa lies buried under more file folders. Paper seeps through the openings of three ancient briefcases held together with tape. "For a man who has made a billion dollars creating media that allow people to communicate without paper, Rogers uses a lot of it," observed Daniel Stoffman, a Toronto writer, in 1989.

In the 1980s, as the relentless pace began to take a toll on Rogers's health, some employees thought they detected a change in their boss. "Ted's easier to work with," said Jim Sward, after Rogers suffered his first heart attack. "He's more considerate, and he deals with issues in more detail. He's made more progress under this new approach. He's basically stopped trying to do everything."

It was, perhaps, wishful thinking on the part of Sward, who left the company without fanfare in 1993 "to pursue other interests in broadcasting," while Rogers kept on ticking at the pace of ten men.

In fact, health concerns have never caused Rogers to slow down. In 1985, he underwent surgery on his eyes for cataracts. For most of his life, Rogers has had only 8 per cent vision in his right eye, and cataracts could have left him blind. The day after the operation, his eyes still covered in bandages, he called a meeting around his bed to discuss a business deal.

In May 1988, within hours of undergoing surgery after a coronary aneurysm, he was up and working in his hospital room. Four years later, he flew to Minnesota for quadruple-bypass surgery at the Mayo Clinic. Before the anesthetic had worn off, he started dictating letters. A few days later, when a reporter asked Cablesystems president Colin Watson when he expected his boss back in the office, Watson said, "What do you mean, 'When?' He's back at work already. I was on the phone with him yesterday for three hours."

A turning point seemed to arrive in October 1993. After announcing that he would sell 10 per cent of his holdings in Rogers Communications for estate-planning purposes, Rogers claimed to be paying more attention to the bottom line and less to the growth of his company. The *Financial Post* suggested that he was turning RCI into an operating company. Rogers attributed the change in focus to the experience of friends who had suffered financial losses at a similar age to his. "I'm frankly very upset about very good friends, close friends, people I respect and admire, having a great many problems," he said. As violins played, reporters wept and investors contemplated their loved ones, Rogers pointed out that the company's chairman, Gar Emerson, had steered Rogers onto a more practical course. "If I spend more time on these things, it sends a signal to people," he said. "I think the emphasis has been on growth and not on operations. With fairness to management, that has been their guideline. The criticism that I make is of myself, not of anybody else."

The message was clear: The decades of rapid growth were over. It was time to focus on running the business at hand and generate a profit. Six months later, Rogers offered $3 billion to buy Maclean Hunter, and the seventy-five-hour work weeks intensified.

"I tend to go from one project to another quickly," Rogers said. "And I get enthusiastic. I do the best I can and I spend long hours at it. My idea of slowing down is putting a fax in the car."

All this effort is made with the aim of winning and dominating his industry. "My personality requires me to be a major factor in what I do," he said in 1988, after selling his U.S. cable holdings for $1.6 billion. "We just weren't ever going to dominate cable in the United States, so I needed a new adventure."

If all this makes Ted Rogers a difficult person to work for, it sometimes makes him impossible to live with. He always wants to have his own way, but he's so god-darned nice about it that you can't disagree with him. It's a style that fatherless sons learn from their mothers at an early age. As a way of dealing with a world that can kill your father before you have time to know him, it can be pretty successful, and pretty disarming, too.

In his personal life, Rogers's wife seems to support his inspired lunacy without much complaint. They met at a party in 1957, when Rogers was staying in Nassau at the house of a fraternity brother. She found him quite shy, and they courted for six years before they were finally married, in September 1963, when Rogers was thirty. Loretta Robinson was the daughter of a British MP, Sir Roland Robinson, who, after thirty-three years, had been rewarded with the governorship of Bermuda and a title, Lord Martonmere, which impressed the colonials to no end. Loretta spent most of her time in the United States. She attended Wellesley College in Massachusetts and obtained a degree in fine arts at the University of Miami. After their marriage, she became a close confidante, adviser and director of Rogers's company. "They have a very consultative relationship," said a friend. "He doesn't make a move without consulting her."

Confidante or not, Loretta has managed to stay in the background during her husband's career, except for a brief and embarrassing moment when she was caught at Toronto's Pearson International Airport returning to Canada on a late-night flight from the Far East with an unset ruby and three sapphires worth $193,000 in her purse. The baubles were a twenty-fifth wedding anniversary gift from her husband and, like thousands of other foolish Canadians who try to sneak past the watchful eye of a customs inspector with bundles of cigarettes and contraband wine stuffed down their trousers, she had tried to evade the duty that the Canadian government exacts from its citizens who think they should operate under their own personal Free Trade Act. For good measure, Loretta also carried a $4,400 diamond-studded necklace, a pair of $5,000 gold cuff-links, a jade pendant, a compact disk player and a camera, all tucked neatly into the folds of her matched set of Louis Vuitton luggage. The practice of smuggling stuff back from Hong Kong is common enough, but the value involved set this case apart from the usual ones that involve Scotch and a couple of CDs. Sensing a juicy story when she saw it, the reporter Linda McQuaig took a break from reading Canada's Income Tax Act and rounded up the usual irrelevant details to paint Loretta Rogers in the worst possible light, itemizing her husband's wealth, her father's title and her family's golf club memberships and adding that if convicted, Loretta might go the Big House for up to five years. Tickled pink at the possibility of a rich man's wife withering away behind bars among a gaggle of tattooed biker chicks and recidivist bunko butches, the *Globe and Mail*'s editors let McQuaig ramble on for a dozen paragraphs. No one paid any attention a couple of years later, when Loretta pleaded guilty to smuggling and was fined $25,000.

The case did no permanent damage to the Rogers children, who have all reached their twenties without shaving their heads and joining a commune, selling drugs to an undercover cop at Harvey's near Allan Gardens, or writing best-sellers about their intimate affairs with the Anglican choirmaster. All born between 1968 and 1972, the kids were used to the unwarranted pressures and late-night conflabs in the den with lawyers bickering over millions of dollars and imminent deadlines for unpaid loans. It came with the territory of growing up in the Rogers family. In fact, Lisa, born in 1968, Edward Samuel III, in 1970, Melinda, in 1971, and Martha, in 1972, all came into the world during one of their father's most difficult periods in business, when he had mortgaged his house and was teetering on the verge of bankruptcy. Rogers tried to avoid talking shop in front of the children. "Are you kidding?" he suggested in mock horror to *Report on Business* magazine's David Olive in 1985. "The things that have happened lately – do I want to scare them out of the business?" But it's hard to ignore your dad when he's watching TV in the bathroom with a toothbrush in his mouth or greeting the president of Maclean Hunter at the front door on a Sunday evening after every newspaper in the country has announced in 48-point type that dad plans to buy the company.

When the children were young, Rogers helped the kids with their homework, played with their dogs, rabbits, budgies and mice, and tried to be the best father he could, since he didn't have one of his own to model himself after. As the owner of a string of radio stations, he could keep his children amused by taking them on tours of CFTR to meet the men behind the demented voices of their favorite disc jockeys. Later, in 1994, after his successful acquisition of Maclean Hunter, Rogers sent his daughter Melinda with a couple of journalists and a photographer from *Maclean's* magazine to interview the prime minister. While Melinda lugged camera bags and tripods around the room, the *Maclean's* editor and Ottawa bureau chief chatted with the prime minister in the customary hard-hitting style that Canadians have come to expect from their national newsweekly. Loath to admit that anyone actually owned their publication, other journalists at the magazine were miffed that they, too, hadn't been allowed to carry some camera bags to Ottawa, and they whined so loudly that another magazine picked up the story.

Behind his family and his business, according to Rogers, lingers the constant spectre of his father, who died when Rogers was five. "I always wanted to be as good as my father was," Rogers has said on many occasions. "The problem is you never feel as if you've really done it, because once you feel you have, you're in trouble."

The elder Rogers was the latest branch on a family tree that extends through Pennsylvania and Massachusetts to England, where one of Rogers's ancestors, a canon of Old St. Paul's in London, was burned at the stake in 1555. A few decades later, Thomas Rogers came to America aboard the *Mayflower*. During the 1700s, the family migrated to Pennsylvania, then did a U-turn and headed to Newmarket, a hotbed of Quakerism north of Toronto, during the American Revolution.

Gradually, the family drifted towards Hogtown, the Big Smoke, growing rich as they bought and sold parcels of land, started businesses and invested their profits. Growing up in Rosedale in the early 1900s, Rogers's father, Edward Samuel Rogers Sr., enjoyed the fruits of his ancestors' industry. Grandfather Sam had made a fortune in oil through his company, Samuel Rogers & Co., which later became part of Imperial Oil. Sam Rogers had also been a co-founder of the Hospital for Sick Children in Toronto. Another relative started Elias Rogers Coal Co., which was purchased in the 1960s by Texaco. An employee of the oil firm recalled watching Sam Rogers borrow $3,000, a considerable amount at the time, with no security but his own good name. In fact, the Rogers clan learned early how to borrow money for their enterprises, a talent that would remain intact within the family and come in handy on countless occasions.

The money itself came and went; the more important legacy

endured in the ability to understand money and put it to productive use. Even if there isn't much around at a particular moment, you can always find a way to raise money if you've grown up in a family that regards it the way a dairy farmer regards his herd, so that you feel comfortable with money and appreciate its value as more than a means for acquiring a hot car, a fancy suit and babes in tight skirts. If you understand how money works, it doesn't intimidate you, so when you find yourself owing $100 million to a bank with one day left before the loan comes due, you don't start to sweat profusely, lose control of your bladder and bark like a dog as you wait in the hallway outside the president's office for a meeting.

The elder Rogers was a driven man and a born tinkerer. At thirteen, he won an award for best amateur-built radio set in Ontario. Preferring his second-floor bedroom-laboratory or the family garage behind their house on Nanton Avenue to more formal surroundings, Ted Sr. dropped out of the University of Toronto's school of practical science to fiddle with radios on his own. In 1921, he was the first amateur radio operator in Canada to transmit a radio signal across the Atlantic.

Rogers's father enjoyed some advantages that he was fated never to share with his son. That year, for example, Ted Sr.'s father bought him the Independent Telephone Company, which manufactured battery-powered crystal radios, complete with head sets, code signals, Nick Danger masks and spinning propellers that you could plug into your Chicago Cubs hat. But with a characteristic streak of obstinacy, Rogers Sr. had other ideas besides running a company that he knew would soon be obsolete, and he developed the technology that would put it out of business. In 1925, in a laboratory on Chestnut Street near Maclean Hunter's headquarters, he perfected a practical version of the alternating current radio tube. Until then, most radios had run on direct current provided by batteries, "huge things that leaked acid over everything," his son said. They consisted of "boxes, batteries, and enough festooned wire to pen a chicken farm," added author Donald Jack. The Rogers tube, however, allowed radios to operate on power supplied by a light socket. Without all the wires and batteries, radios quickly evolved into home appliances, and it wasn't long before every housewife in America was singing along to the Mills Brothers as she peeled the carrots at the kitchen sink.

Like his son, Ted Sr. put in long days and nights perfecting his equipment and correcting mistakes. These became more urgent after a supplier, affiliated with a rival radio manufacturer, began selling defective filaments to Rogers for his tubes in the hope of driving him out of business. "Dad was not an idea genius," said Rogers. "He spent years patiently trying different combinations to make AC electricity work radios."

Financed primarily by his father, Rogers Sr. started Rogers Radio Tube Co. to make the tubes. He also started Standard Radio Manufacturing Co., which incorporated the tubes into the Rogers "Majestic," the most advanced plug-in radio of its time. The first Rogers Majestic went on display in 1925 at the Canadian National Exhibition.

Until Ted Rogers Sr. came along, radio delivered programs that sounded like a Sunday afternoon get-together with the aunts playing piano and the uncles swapping stories on the back porch. Programs often arrived in people's homes over the airwaves in dismal condition, after being mugged and scrambled by interfering signals from the United States. To eliminate this problem and give people something better to listen to on the radio that he was trying to sell, Rogers Sr. upgraded his ham radio station, 9RB, to commercial status and called it CFRB – Canada's First Rogers Batteryless. He was then twenty-seven years old.

CFRB was the sixth commercial radio station in Toronto. The first, called CFCA, was owned by the Toronto Star and operated by a dorky kid with a squeaky voice named Foster Hewitt, the son of the paper's sports editor. Rogers's radio station was different. CFRB was the first radio station in the world to transmit signals using batteryless transmission; it broadcast sound of a far superior quality to its rivals.

In 1931, Rogers Sr. acquired one of the first four licences in Canada for experimental TV broadcasting. By 1938, he had started working on radar development. But the next year, he died suddenly of an aneurysm, leaving his wife with little financial support and his five-year-old son with a competitor whom he could never hope to beat, even if he spent a lifetime trying. As a child, Rogers went to the CNE and remembers seeing radio sets with his father's name on them. By then Rogers Majestic belonged to Philips, which had bought the company three months after Ted Sr. died. Over the following years, Rogers's mother, Velma, and other heirs to the family fortune also sold the family shares in Standard Broadcasting Corp., which owns CFRB.

"He didn't have much life insurance," said Rogers, who still wears his father's gold ham radio ring, bearing the call number 3BP, and keeps his father's radio tubes in a glass-and-mahogany case in his den, under a portrait of Ted Sr. "All I've got from my father is a Valentine's card. He just worked himself to death. He was my hero and role model, and I think I was very motivated by that, because I felt cheated that he died. And later I felt cheated that the businesses were all sold.

"If you have a feeling you've been robbed, you have a touch of bitterness and it really propels you. It's also very hard to compete with your father. It has always driven me."

Rogers inherited a few vague memories and a superhuman legend to remind him of the man. He recalls that his father was intensely shy and uncomfortable in the presence of people he didn't know well. His enthusiasm was infectious, Rogers said, but he put most of his time, thought and energy into his work. "Whatever he did, he did real hard," Rogers told Donald Jack in *Sinc, Betty and the Morning Man.* "He often worked all night in the tube plant. By the time he was thirty-nine [sic], the year of his death, he had burned himself out."

But the intensity lived on in his son. "As I grew up, I was filled with all the stories of my dad, the development of communications, and his part in it – I was imbued with the excitement of it all."

The memory of his father became a touchstone for his efforts and a sounding board for his thoughts. "My mom told me that Dad's invention was the result of constant application of different circuit combinations until he got the right idea," Rogers said. "Sometimes people lose things for short-term reasons, but that's not good. A constant effort should be made for a long-term goal."

Although he refers constantly to his father, the dominant force in his family after 1939 was his mother. For thirty-three years, until her death in 1971, she provided him with advice and support about his life and his business. In her final days, the directors of Rogers Cable held board meetings around her bed. "My mother was sort of a great hero, my pal, and partner," he said. "She was sick for a long time, and I think she taught me a great deal. Not to give in, not to give up."

Two years after his father died, Velma was remarried to John Graham, a partner with Blake Cassels & Graydon and an Old Boy of Upper Canada College, class of 1930. Graham's best man at the wedding was Don Hunter, son of the founder of Maclean Hunter. "We would visit the Hunters' cottage a number of times every summer," Rogers said. "Later, we used to trade ten-year budgets across our adjoining back fences. I talked him into buying CKEY. I sold his house to another friend when he bought a new one down the road. We were good pals."

For a brief time, Graham provided Rogers with fatherly influence and advice, and he resumed this role when Rogers started in business. "When I'd say, 'Let's do it now,' he'd say, 'Let's sleep on it,' " Rogers recalled. In the meantime, though, as Canada sent troops to fight the Second World War, Graham went overseas, and Rogers went to UCC.

As a benefit of a privileged childhood, UCC left something to be desired. "I used to get caned quite a bit," Rogers said. "It was like being in jail."

At times, the school served as a surrogate father and perhaps saved Rogers from turning into the sissy or the psychopath that he might have become if he'd stayed home with his mother and a nanny knitting tea cosies and rearranging the furniture. It also provided him with ample time to tinker with radios, whether or not such an activity was officially sanctioned by the grey-haired ramrods who ran the place. "I used to sneak in crystal sets and radios when I was boarding in prep," he told James Fitzgerald in *Old Boys: The Powerful Legacy of Upper Canada College*. "I put an antenna like a clothesline across [prep headmaster] Alan Stephen's garden. I didn't think he'd notice, but he did. So they moved me away from a window cubicle to the other side of the dorm. Then when everybody was away I put the wires underneath the linoleum. The linoleum ripped, so I got caned again."

Rogers eventually distinguished himself at UCC as such a goof-up that his housemaster made him sleep in the room next to his office. From there, at fourteen, Rogers rigged up a TV antenna on the roof. The apparatus was attached to a wire so that Rogers could raise it to receive signals from Buffalo. Rogers would unlock a closet, and he and the prefects would gather round a TV set, which he kept inside. When the program was over, he'd put the TV away, lock the closet and lower the antenna so that no one could see it from the ground. One icy winter night, the antenna froze to the roof. Struggling to raise the thing, Rogers pulled so hard on the wire that the antenna ripped out of its moorings, toppled off the roof, through a window and into the basement. The future investment bankers, who enjoyed Rogers's company as long as he provided the entertainment, headed for cover down the hall, leaving Rogers to take the heat. "I had to take the rap – six of the best," he said. "But I still like to say it was the first Rogers TV cable system in Canada."

Rogers's quirkiness didn't sit well with the future honchos who threw their weight around the corridors of UCC. "He was always a singular person," recalled David Scott, a schoolmate, who was interviewed by the Vancouver *Sun* in 1989. "There was a lot of peer pressure to change his views, and sometimes he was bullied. The thing I remember about Ted Rogers was that even though bullied and under tremendous pressure to follow along, he refused to do so."

In his final year, Rogers lived at home, taking a limousine to school in the morning and driving home in the afternoon in his sports car, which was left for him each day outside his classroom. "He was a rebel," said Joe Armstrong, a Toronto writer. "It's funny to see him now. He looks magnificently respectable. But he was a rebel in private school, where the idea is to make you conform to the Wood Gundy mould."

Looking back on his adolescent experience, Rogers concluded, "Everybody has their share of unhappiness, and I certainly had mine."

From UCC, Rogers went down the hill to the University of Toronto, where he joined many of his high-school classmates at Trinity College to study political science and economics. Like UCC, Trinity maintained high academic standards along with a lot of dotty traditions like wearing gowns to dinner, which distinguished its students from others at the university, allowed them to feel superior and made them look like dorks. At university, Rogers met George Fierheller and became involved in student politics. Disenchanted with conventional student political groups, Rogers formed his own, the Independent Political Association. Taking its initials from India Pale Ale, a popular beer, the party captured eight seats on the student council. In victory, Rogers shook hands with his opponent and said, "I know we've been fighting, but why don't we be friends and work together," a line for which many a corporate president would fall over the years as Rogers took over his company.

Later, Rogers joined the Progressive Conservative Student Federation, becoming its president for a time until he was removed for spending its entire budget on a lunch for John Diefenbaker. He then formed Youth for Diefenbaker and travelled around the country signing up student supporters for the future prime minister. One of the supporters he attracted on a visit to the Maritimes was another future prime minister, Brian Mulroney. In an interview with Lawrence Surtees, George Fierheller recalled that Rogers's idea of a good time "was running around Forest Hill pulling up lawn signs of rival political parties." Entering the United States during the McCarthy years, Rogers announced to an immigration inspector that he was a *Progressive* Conservative. The inspector detained him.

Taking a break from this non-stop hilarity, Rogers went to Ottawa in his final year at Trinity on behalf of the National Progressive Conservative Student Federation to present a brief to the Fowler royal commission on broadcasting. The commission's legal counsel was Jean de Grandpré, who had such an affinity for the regulatory hokey-pokey that he would later become chairman of BCE.

As a high-school student on summer vacation in Muskoka, Rogers had become involved in booking dance bands at local resorts. At university, he became a full-fledged booking agent, signing bands at clubs and dance halls around the city and undercutting the competition for gigs with union stalwarts in the business like Moe Kauffman. One New Year's Eve, Rogers found himself setting up ten bands in ten different locations around the university and selling photographs of undergrads and their dates at each event. "I'd get on the PA and pitch the guys on buying a picture of the girl of their dreams," he said. "I'd charge them $2.50, but I paid only 47¢ for the components. I made a nice profit," he said, "with very little capital expenditure."

By the time he entered law school at Osgoode Hall, on the urging of his mother, he needed an activity more engrossing than booking dance bands to keep him occupied. "I wanted to get into radio," he said. "The question was how." It didn't take him long to find an answer. In 1957, he bought a station.

With his dance bands and radio stations, Rogers gained his first taste of show business. Among other things, it taught him to appreciate the value of first impressions, a lesson he has put to good use in more recent years, especially with journalists. Knowing that the media feed on gossip, he's not about to give them more than he has to. He says nothing in public that he doesn't want anyone else to hear, and he often says the same thing, over and over again, in a routine that he has followed for more than fifteen years.

First he disarms the journalist in his downtown Toronto office, which, as the journalist points out without fail, affords a "spectacular," "dazzling," "panoramic" view of the lake. If all goes smoothly and the ink-stained wretch doesn't put his desert boots on the coffee table or sneeze into the sugar bowl, Rogers invites him to another less formal meeting at his "palatial," "fashionable," "fifteen-room" house in Forest Hill. Dressed in tennis sweater, cotton slacks and running shoes that squeak as he crosses the marble floor of the vestibule, Rogers greets the rube at the front door and leads him to the den, past a painting by Loretta that hangs above the drinks tray outside the dining room, where a dining-room suite "reigns with a grandeur reminiscent of Versailles." In the "book-lined," "mahogany-panelled" den, under the portrait of his father, Rogers idly pats the family dog as he delivers the inside dope about life, business and his inner feelings on the human predicament. Whether it's because of Ted Rogers's charm or the potency of the wine he serves his guests, the reporter ends up writing stories that make Rogers sound like the Dalai Lama of telecommunications. "Lean, silvery, quixotic," said David Smith about Rogers after interviewing him for the Vancouver *Sun*. "Tall, slender, fair and beaming," gushed Judy Steed in the *Toronto Star*.

Some lucky journalists get to take the third step in the program and visit Rogers in the Bahamas. At that point, the writer blows a head gasket and the purple prose flows like a light wave down a fibre-optic cable. "An antebellum mansion on the lush island of New Providence in the azure seas of the Bahamas," gushed fellow UCC alumnus Peter Newman, clad for the occasion in verbal pith helmet and journalistic Tilley suit, as his glass of ice water grew "tepid as we talked into the tropical afternoon." The Rogers home is "a three-minute golf cart ride from the legendary Lyford Cay Club that provides a safe roost for tax-weary millionaires on the wing," continues Newman. "The house resembles a movie set version of a colonial governor's chancellery, complete with elegantly rotating overhead fans and the aura of power calmly possessed and habitually exercised." After this lush ditty from the Liberace of the printed word, readers can hardly expect anything particularly controversial, and Newman lives up to their expectations.

The odd journalist who doesn't get himself invited home ends up sulking and lobbing elegantly rotating but ineffective verbal grenades at Rogers. One disgruntled scribe with a chip on his shoulder the size of a small planet thought he would really expose once and for all the underbelly of Canada's communications czar when he wrote with barely restrained glee in *Canadian Business*: "Most entrepreneurs don't have the financial resources or web of connections that Rogers enjoys." In case readers didn't get the message, he repeated his damning observation a few hundred words later. "But his success is also built on a cunning way with regulators and connections enjoyed by few others."

In fact, connections have certainly helped Rogers, just as they probably helped the writer get his own job. Rogers's connections helped in the 1970s, when his friends let him extend the deadline on a $2.5-million payment for their share of his company. At other times, they might have helped him get in the door. But ultimately, Rogers has had to pay his debts and get the job done, just like the rest of us. There are no free rides, and bankers are your friends only as long as you pay your debts. Far more helpful to Rogers than his connections was his upbringing, and that's attributable to nothing more nefarious, reprehensible or cunning than plain dumb luck. If life is what you make of it, Rogers can't be faulted, although in a spirit of fairness, the *Canadian Business* reporter kept trying. He noted finally that Rogers has made a lot of money on paper, adding pointedly, "and he intends to keep it for himself."

To his credit, Rogers hasn't kept all his money for himself. While the most generous benefactors are the ones you never hear about, Rogers hasn't drawn much attention to his own generosity. In 1989, he gave $2 million to the Rogers Communications Centre at Ryerson Polytechnic University in Toronto, beating Bell Canada to the punch by six years. With state-of-the-art facilities, the $25-million Rogers Centre provides industry-standard communications facilities for students and faculty involved in journalism, radio and television arts and computer science. The radio station CJRT broadcasts from studios in the centre. With $500,000 in equipment donated by AT&T, the centre also supports the development and testing of interactive communication systems and the training of students in up-to-date technologies. Coincidentally, Loretta Rogers sits on Ryerson's board of governors. So does their neighbour, Isabel Bassett. In 1995, Centennial College opened the Bell Centre for Creative Communications a few miles away in East York.

As usual, however, Rogers drew more criticism than Bell for his gesture, perhaps because, unlike Alexander Graham Bell, he's still alive and makes an irresistible moving target. Nor does anyone named Bell have to announce publicly his annual income from the company that bears his name.

Offended that a financial donor could buy his way into their community, some faculty members and students at Ryerson immediately took umbrage over the name of the Rogers Centre. "I object to the smarmy way the new building at Church and Gould has been named the Rogers Communications Centre," wrote Loren Lind, an instructor of journalism at Ryerson and an eloquent defender of the downtrodden. "So Rogers gave $2 million. That's nice," Lind ruminated. "For that Ted Rogers deserves a plaque on the wall, but hardly the name of the whole thing."

Continuing this penetrating analysis of Canadian culture for another two paragraphs, Lind pointed out that the people of Ontario, through their taxes, had worked their fingers to the collective bone to contribute $1.75 apiece towards the building, while Rogers had simply pulled out his wallet and handed over $2 million. The unassuming taxpayers who had worked so hard for so little would never expect the building to carry all 9 million of their names, so why should Rogers? Isn't that typically Canadian? Lind concluded. "It will be a constant reminder of how things work in our country: the rich get the credit by heaping a million or two on top of the millions and millions that represent the sweat of all the rest of us." In fact, the centre is named after Rogers's father, although that's not readily apparent since they both have the same name.

On a more human scale, when Rogers Cablesystems stopped providing CNN on a free trial basis during the Gulf War, a subscriber with relatives in Israel phoned the company to demand resumption of the service. A Rogers sales rep said it would take a week to reconnect the subscriber. So the subscriber contacted Rogers at home by phone on a Saturday afternoon. The next morning, Rogers ordered a service truck to visit the customer's house and install his CNN service.

Still, Rogers hasn't given everything away, and the question remains what will happen to it all when he's gone. Within the family, Rogers started planning his estate when he was thirty-five years old and an insurance salesman's dream. "In addition to lots and lots of life insurance, I've planned for an organization to continue when I die," he said in 1981. "So from a family standpoint, my investment is fine.

At the corporate level, his employment contract calls for him to step down at the end of the year 2000. He will then work on a consulting contract for $150,000 a year for five years. By then he'll be seventy-five, ready for a dip in the pool.

Regarding his successor, Rogers has not said who will hold voting control over his estate, although he has said none of his children will take over immediately. "I have four children," he said in 1994, "and I have no idea whether they all have the interest or the capability or the stamina to be able to do the things you need to do."

His only son, Edward, has been working for several years at Comcast in Philadelphia, the company that bought Maclean Hunter's U.S. cable systems. His daughter Lisa works for Rogers Cablesystems. His third daughter Melinda has also worked for Rogers. "Of course I would like Edward to do this job one day," he said in a comment that feminists have yet to hold against him. "But under what I've set up, he will only end up in this office if he really deserves and wants to be here."

In the meantime, the company will not be sold, if Rogers estate works according to plan. "If anyone tries to put their hand to a pen to sign to sell the business, I'll rise up and kill 'em!" he said.

In his own mind, that day lies far in the future. The man whose accomplishments, according to one cable executive, "make the rest of us look absolutely pygmy-like"; who pioneered FM radio, cable and cellular telephone in Canada; who's regarded as a "crazy, crazy, crazy lunatic"; who saw the future twenty years ago – this man still has a way to go. "He's surprised by the way people regard him," said Graham Savage. "Ted Rogers still thinks of himself as a young man who's establishing himself."

Bibliography

Books

Ellis, David. *Home Entertainment & the New Technologies*. Toronto: Friends of
    Canadian Broadcasting, 1992.

Fitzgerald, James. *Old Boys: The Powerful Legacy of Upper Canada College*.
    Toronto: Macfarlane Walter & Ross, 1994.

Haslam, Fred. *A Record of Experiences with and on Behalf of the Religious Society of
    Friends in Canada with the Canadian Ecumenical Movement 1921-1967 with
    Some Thoughts to the Future*, 1968.

Jack, Donald. *Sinc, Betty & the Morning Man: The Story of CFRB*. Toronto:
    Macmillan, 1977.

Mirabito, Michael, and Barbara Morgenstern. *The New Communications
    Technologies*. Boston: Focal Press, 1990.

Surtees, Lawrence. *Wire Wars: The Canadian Fight for Competition in
    Telecommunications*. Toronto: Prentice Hall, 1995.

Magazines

Brown, Ian. "The Hoisting of the Jolly Rogers." *Maclean's* 22 Jan. 1979.

Burrows, Peter, "Pick of the Litter: Why SBC Is the Baby Bell to Beat." *Business Week*
    6 Mar. 1995.

Dalglish, Brenda. "King of the Road." *Maclean's* 21 Mar. 1994.

———. "The King of Cable." *Maclean's* 22 Mar. 1993.

Daly, John. "The Next Hurdles." *Maclean's* 21 Mar. 1994.

"Discoveries & Inventions: Communications." *Horizon Canada* 4 (47) 1986.

Edmonton, Gail, Karen Lowry Miller, Julia Flynn, and Mark Lewyn. "Bonn's
    Telecom Bombshell." *Business Week* 13 Feb. 1995.

Fisher, Ross. "Ted's Team." *Canadian Business* Aug. 1989.

Grover, Ronald. "Cable Operators Bite Back." *Business Week* 13 Mar. 1995

Hawkins, Chuck. "A Cable Mogul with a Live-Wire Idea." *Business Week* 4 July 1988.

Johnson, Phil. "Rogers: The Empire Strikes Back." *Broadcaster* Aug. 1986.

Kessler, Andrew J. "The Phone Companies' Dirty Secret." *Forbes* 10 Apr. 1995.

Lilley, Wayne. "Outstanding CEOs of 1981." *Canadian Business* July 1982.

Marion, Larry. "The Legacy." *Forbes* 6 July 1981.

McElgunn, Jim. "Rogers Blasts Phone Companies at Cable Conference." *Marketing* 23 May 1994.

McGugan, Ian. "Such Good Friends." *Canadian Business* Apr. 1994.

McMurdy, Deidre. "The Rules of the Game." *Maclean's* 21 Mar. 1994.

———. "The Colonel's Legacy." *Maclean's* 21 Mar. 1994.

Meeks, Fleming, with Roula Khalaf. "This Will Be a Very Political Issue." *Forbes* 19 Feb. 1990.

Misutka, Frances. "Romancing the Phone." *Canadian Business* Nov. 1992.

Munk, Nina. "Culture Cops." *Forbes* 27 Mar. 1995.

———. "Ted Rogers' New Apartment." *Forbes* 24 Apr. 1995.

Newman, Peter C. "Let the Second Force Be with You." *Maclean's* 20 Mar. 1989.

———. "Life in the Fast Lane." *Maclean's* 21 Mar. 1994.

———. "The Ties that Bind." *Maclean's* 21 Feb. 1994.

Olive, David. "Newsmakers of 1991." *Report on Business* Jan. 1991.

———. "The High-Wire Act of Ted Rogers." *Report on Business* July/Aug. 1985.

"Our Annual Ranking of Corporate Power Brokers." *Canadian Business* Nov. 1989.

"Rogers Adds Baton Shares to Portfolio." *Broadcaster* Sept. 1993.

Samuels, Gary. "Lord, Make Me Competitive – But Not Just Yet." *Forbes* 13 Mar. 1995.

Schine, Eric, and Kathleen Kerwin. "Digital TV: Advantage, Hughes." *Business Week* 13 Mar. 1995.

Stoffman, Daniel. "Great Connections." *Report on Business* Sept. 1989.

"The Legacy of Edward Rogers Sr." *Report on Business* July/Aug. 1985.

"The Winner's Circle." *Broadcaster* Mar. 1994.

Newspapers

Adolph, Carolyn. "Rogers Buys into CNCP in Long-Distance Challenge." *Toronto Star* 20 Apr. 1989.

Barnes, Al. "Drive to Succeed, Father's Influence Make Rogers Run." *Toronto Star* 3 Feb. 1994.

Bell, Andrew. "Rogers' Decision Spares It More Debt." *Globe and Mail* 20 Apr. 1995.

Best, Dunnery. "Watch Rogers Flow in Stronach's Footsteps." *Globe and Mail* 13 May 1995.

Best, Patricia. "Mr. Rogers' Neighborhood." *Financial Times of Canada* 5 June 1989.

Blackwell, Richard. "Rogers Says He Welcomes Competition." *Financial Post*
    21 Sept. 1989.

Brehl, Robert. "Avoid 'Brawl,' Rogers Pleads." *Toronto Star* 2 Mar. 1995.

———. "Cabinet Demands CRTC Make Cable Access 'Fair.'" *Toronto Star* 15 Mar. 1995.

———. "Cable Firms Fit to Take on Rivals, Phone Study Says." *Toronto Star* 16 Feb.
    1995.

———. "Hidden Cable 'Taxes' Blasted." *Toronto Star* 30 Mar. 1995.

———. "Rogers Leaves Unitel Hanging." *Toronto Star* 20 Apr. 1995.

———. "Unitel Admits Error in Access Payments." *Toronto Star* 6 Apr. 1995.

Bruce, Alexander. "The Lines are Drawn for a Rogers Tel." *Financial Times of
    Canada* 3 Apr. 1989.

Campbell, Murray. "Bell's Plan to Trim Thousands of Staff Rings True to Analysts."
    Globe and Mail 29 Mar. 1995.

Chevreau, Jonathan. "High-Tech Pioneers Converge in Toronto." *Financial Post* 14
    July 1994.

———. "Ted Rogers Steps Down from Board of Unitel." *Financial Post* 30 June 1994.

Cocoran, Terence. "Satellite Competition Should Not Be Delayed." *Globe and
    Mail.* 7 Apr. 1995.

———. "The Telecom Wars Have Only Begun." *Globe and Mail* 22 Apr. 1995.

Cook, Peter. "Why Battle Canada's Big Corporations?" *Globe and Mail* 22 Sept.
    1994.

Critchley, Barry, and Susan Griffiths. "Jolly Rogers." *Financial Post* 31 May 1991.

Enchin, Harvey. "Cogeco Calls for Stentor Breakup." *Globe and Mail* 24 Mar. 1995.

———. "It's Official: CRTC Okays MH Takeover." *Globe and Mail* 20 Dec. 1994.

———. "Numbers Didn't Add Up for Bell." *Globe and Mail* 21 Apr. 1995.

———. "Ottawa Seeks Rules on Access to Cable TV." *Globe and Mail* 15 Mar. 1995.

———. "Rates Out of Line, Rogers Told." *Globe and Mail* 22 Sept. 1994.

———. "Rogers Hedges on Cheaper Rates." *Globe and Mail* 21 Sept. 1994.

———. "Rogers Makes Cross-Canada Pitch." *Globe and Mail* 29 Apr. 1992.

———. "Rogers Makes New Pitch in Court." *Globe and Mail* 28 Feb. 1994.

———. "Rogers Shares Sink to a 52-Week Low." *Globe and Mail* 6 May 1995.

———. "Rogers Team Ready to Prove Big Is Better." *Globe and Mail* 20 Sept. 1994.

Enchin, Harvey, and Barrie McKenna. "Rivals Watching Troubled Unitel." *Globe
    and Mail* 20 Apr. 1995.

Ferguson, Jonathan. "Rogers Bid Faces Probe." *Toronto Star* 3 Feb. 1994.

Flavelle, Dana. "AT&T Man Takes Unitel Helm." *Toronto Star* 9 Oct. 1993.

———. "Bill Gates Has Seen the Future." *Toronto Star* 20 Aug. 1993.

——. "Rogers Bids $2.8 Billion." *Toronto Star* 12 Feb. 1994.

——. "Rogers Makes $3 Billion Bid for Cable Rival." *Toronto Star* 3 Feb. 1994.

Gibbens, Robert. "Ted Rogers Makes Unity Pitch." *Financial Post* 29 Apr. 1992.

Gluskin, Ira. "Ted Rogers Walks the U.S. Tightrope." *Financial Times of Canada* 27 June 1988.

Goold, Douglas. "Big Guns Split on Rogers' Takeover." *Globe and Mail* 16 Feb. 1994.

——. "Money Managers Give Thumbs Up." *Globe and Mail* 9 Mar. 1994.

——. "Ted Rogers' Burnt Offering to King Spicer." *Globe and Mail* 8 Sept. 1994.

Haliechuk, Rick. "Ted Rogers Got $610,000 in '93." *Toronto Star* 12 Feb. 1994.

Heinzl, Mark. "Ted Rogers' Bid Would Fortigy Cable Empire." *Wall Street Journal* 4 Feb. 1994.

Hubbard, Jaimie. "Ted Rogers Uses Stock Slide to Buy Bigger Chunk of RCI." *Financial Post* 19/21 May 1990.

——. "Tough, Brilliant Rogers Quests for Birthright." *Financial Post* 24 Apr. 1989.

Jorgensen, Bud. "A Bad Day for Maclean Hunter." *Globe and Mail* 23 Feb. 1994.

——. "Cantel's Share Issue That Never Was Had Its Funny and Not-So-Funny Sides." *Globe and Mail* 1 Mar. 1989.

——. "Revised Offer Fair, Bay Street Indicates." *Globe and Mail* 9 Mar. 1994.

——. "Rogers Increasing His Clout in Video Market." *Globe and Mail* 22 Sept. 1989.

Kelly, Doug. "Rogers Deal Reviewed." *Financial Post* 15 Jan. 1992.

Marotte, Bertrand. "'Visionary' Has Seen Broadcast Battles." Vancouver *Sun* 4 Feb. 1994.

——. "Communication Czar Starts New Chapter." Calgary *Herald* 9 Feb. 1994.

——. "Maclean Hunter Bid Set at $2.8 Billion." Montreal *Gazette* 12 Feb. 1994.

——. "Rogers Puts $2.8 Billion on Table for MH." Vancouver *Sun.* 12 Feb. 1994.

——. "Ted Rogers: Driven Hard by a Dream." Montreal *Gazette* 4 Feb. 1994.

Mayers, Adam. "Why Ted Rogers Should Be Afraid." *Toronto Star* 20 Mar. 1995.

McKenna, Barrie. "Merger May Face Long Regulatory Scrutiny." *Globe and Mail* 9 Mar. 1994.

——. "Rogers' Priority is Long-Distance Licence." *Globe and Mail* 27 Feb. 1990.

——. "Wings Plucked from MH Butterfly." *Globe and Mail* 23 Feb. 1994.

McQuaig, Linda. "Wife of Cable TV Magnate Facing Customs Charges." *Globe and Mail* 1 Sept. 1995.

Miller, Brian. "Rogers Likes to Dance, but 'Hard to Crash Party.'" *Globe and Mail* 25 Feb. 1988.

Papp, Leslie. "Hydro to Study Data Transmission Options." *Toronto Star* 13 Mar. 1995.

Partridge, John. "$75-Million in Debt Motivated Rogers." *Globe and Mail* 15 Jan. 1992.

——. "A $200-Million Jaunt." *Globe and Mail* 17 Jan. 1992.

——. "Bay Street Points to Rogers after Large WIC Purchases." *Globe and Mail* 3 Mar. 1989.

——. "Moffat Talks May Limit Rogers Stake." *Globe and Mail* 18 Aug. 1988.

——. "Rogers Buying 40 Per Cent of CNCP, Planning to Tackle Phone Monopolies." *Globe and Mail* 20 Apr. 1989.

——. "Rogers Confirms He Owns Block of Shares in WIC." *Globe and Mail* 21 Mar. 1989.

——. "Rogers Denies Profiting on Knowledge of Share Issue." *Globe and Mail* 14 Jan. 1992.

——. "Rogers Eyed Paging Unit First." *Globe and Mail* 14 Mar. 1994.

——. "Rogers Finds His Rivals More Reasons to Worry." *Globe and Mail* 3 Apr. 1989.

——. "Rogers' Growth Troubles Torstar." *Globe and Mail* 8 Mar. 1994.

——. "Rogers Likely to Extend Bid." *Globe and Mail* 7 Mar. 1994.

——. "Rogers Loses Court Skirmish." *Globe and Mail* 1 Mar. 1994.

——. "Rogers Officially Drops Its Selkirk Takeover Bid after Southam Opposition." *Globe and Mail* 24 Oct. 1987.

——. "Rogers Picks Juneau to Oversee MH." *Globe and Mail* 3 Mar. 1994.

——. "Rogers Renews Western Cablevision Bid." *Globe and Mail* 7 Feb. 1989.

——. "Rogers Tackles Its Foreign Ownership Problem." *Globe and Mail* 11 Nov. 1987.

——. "Rogers to Undergo Heart Surgery." *Globe and Mail* 13 July 1992.

——. "Sale of Moffat Stake Proves Profitable for Rogers." *Globe and Mail* 7 June 1989.

——. "Ted Rogers Dropping Bid to Increase Moffat Stake." *Globe and Mail* 6 Oct. 1988.

——. "TSE to Decide Who Should Have Disclosed Moffat Deal." *Globe and Mail* 16 Aug. 1988.

Partridge, John, and Harvey Enchin. "Maclean Hunter Better Dead than Alive." *Globe and Mail* 22 Feb. 1994.

——. "Maclean Hunter Rejects Rogers." *Globe and Mail* 16 Feb. 1994.

——. "MH Bid Likely to Dominate Annual Meeting." *Globe and Mail* 8 Mar. 1994.

——. "Rogers, Maclean Hunter Sign Deal." *Globe and Mail* 9 Mar. 1994.

——. "Rogers Tempers His Takeover Stand." *Globe and Mail* 25 Feb. 1994.

——. "Rogers Was the Only Game in Town." *Globe and Mail* 9 Mar. 1994.

Perry, Robert. "Pay-TV Monitors Ways to Fend off Fade to Black." *Financial Post* 4 Feb. 1984.

Rinehart, Dianne. "Consumers Should Underwrite Fight with Deathstars: Cable

Magnate." Montreal *Gazette* 5 Mar. 1993.

Saunders, John. "Rivals Circle in on MH Units." *Globe and Mail* 9 Mar. 1994.

Siklos, Richard. "Rogers Blasts Colleagues." *Financial Post* 22 Oct. 1991.

———. "Rogers on Track with MH Selloff." *Financial Post* 21 June 1994.

———. "Rogers Vows to Shake Up CTV." *Financial Post* 21 June 1994.

———. "Ted Rogers Buys 8% Stake in Baton." *Financial Post* 27 Aug. 1993.

———. "Ted Rogers Hasn't Stopped Growing Yet." *Financial Post* 29 Oct. 1992.

———. "The Wide, Wired World of Ted Rogers." *Financial Post* 2 Oct. 1993.

Smith, David. "The King of Cable." Vancouver *Sun* 3 June 1989.

Spears, John. "Buyout Bid 'Reasonable,' Fund Man Says." *Toronto Star* 12 Feb. 1994.

Steed, Judy. "A Cable King's Field of Dreams." *Toronto Star* 13 Mar. 1994.

———. "Ted's Excellent Adventure." *Toronto Star* 13 Mar. 1994.

Surtees, Lawrence. "Bill and Ted's Adventure Will Be a Tough Journey." *Globe and Mail* 15 Jun. 1992.

———. "CRTC Inquiry to Shape Canada's Broadcast Future." *Globe and Mail* 6 Mar. 1995.

———. "More Room at the Top at Rogers." *Globe and Mail* 1 Apr. 1993.

———. "Rogers in for the Long Haul with Unitel." *Globe and Mail* 23 Apr. 1991.

———. "Rogers Reviews Unitel Stake." *Globe and Mail* 20 Dec. 1994.

———. "Stentor Slams Long-Distance Rivals." *Globe and Mail* 21 Mar. 1995.

———. "The Battle to Save Unitel." *Globe and Mail* 21 Mar. 1995.

———. "Allegations Fly as Head of CBTA Resigns." *Globe and Mail* 18 Mar. 1995.

Urlocker, Mike. "Rogers Gets Grilling Over Phone Ambitions." *Financial Post* 23 Apr. 1991.

———. "Unitel's CRTC Bid 'A Step Forward.'" *Financial Post* 19/21 May 1990.

Velocci, Randy. "Ted Rogers: In the Spotlight." *Globe and Mail* 8 Feb. 1994.

Winsor, Hugh. "CRTC Considers Ottawa's Proposals on Satellite-TV." *Globe and Mail* 20 Apr. 1995.

———. "Many Strings Pulled in Satellite-TV Show." *Globe and Mail* 1 May 1995.

Yakabuski, Konrad. "Cable TV Czar May Carve up Empire." *Toronto Star* 12 Feb. 1994.

Zerbiasias, Antonia. "Whose Side Is Keith Spicer on?" *Toronto Star* 19 Feb. 1995.

Index